A MODERN HISTORY OF THE STOMACH: GASTRIC ILLNESS, MEDICINE AND BRITISH SOCIETY, 1800–1950

Studies for the Society for the Social History of Medicine

Series Editors: *David Cantor*
Keir Waddington

Titles in this Series

1 Meat, Medicine and Human Health in the Twentieth Century
David Cantor, Christian Bonah and Matthias Dörries (eds)

2 Locating Health: Historical and Anthropological Investigations of Place and Health
Erika Dyck and Christopher Fletcher (eds)

3. Medicine in the Remote and Rural North
J. T. H. Connor and Stephan Curtis (eds)

Forthcoming Titles

War and the Militarization of British Army Medicine, 1793–1830
Catherine Kelly

A MODERN HISTORY OF THE STOMACH: GASTRIC ILLNESS, MEDICINE AND BRITISH SOCIETY, 1800–1950

BY

Ian Miller

Routledge
Taylor & Francis Group
LONDON AND NEW YORK

First published 2011 by Pickering & Chatto (Publishers) Limited

Published 2016 by Routledge
2 Park Square, Milton Park, Abingdon, Oxfordshire OX14 4RN
711 Third Avenue, New York, NY 10017, USA

First issued in paperback 2015

Routledge is an imprint of the Taylor & Francis Group, an informa business

BRITISH LIBRARY CATALOGUING IN PUBLICATION DATA

Miller, Ian.
A modern history of the stomach: gastric illness, medicine and British society, 1800–1950. – (Studies for the Society for the Social History of Medicine)
1. Stomach – Diseases – Treatment – Great Britain – History – 19th century.
2. Stomach – Diseases – Treatment – Great Britain – History – 20th century.
3. Social medicine – Great Britain – History – 19th century. 4. Social medicine – Great Britain – History – 20th century.
I. Title II. Series
362.1'9633'00941-dc22

ISBN-13: 978-1-138-66450-0 (pbk)
ISBN-13: 978-1-8489-3181-7 (hbk)

Typeset by Pickering & Chatto (Publishers) Limited

CONTENTS

For Alice

ACKNOWLEDGEMENTS

This project was kindly funded by the Bardhan Research and Education Trust (BRET), and stemmed from three years of PhD research at the Centre for the History of Science, Technology and Medicine (CHSTM), University of Manchester. I owe a very special debt to Ian Burney, who provided in-depth academic advice and help. I wish to thank Ian in particular for his endless patience and time spent reading through various drafts of this project, and his support for my career following the completion of the project. Ian's input was invaluable at every stage throughout the writing and research for this book.

The initial idea for the project came from Chandu Bardhan, whose support and enthusiasm for my findings (many of which diverged completely from the project's initial aims) I am profoundly grateful for. His input and comments from the perspective of a practising gastroenterologist proved to be invaluable throughout all stages of the project. He was also responsible for obtaining the funding for the project, which was supported by BRET. I also wish to thank the staff at BRET in Rotherham Hospital for their support throughout the writing of this project.

Close academic guidance was also provided by Michael Worboys and John Pickstone during my time at CHSTM. I am also indebted to Rhodri Hayward, Frank Tovey and Hugh Baron for their feedback, as well as to Julie-Marie Strange at the University of Manchester who initially encouraged me to pursue the project. Roberta Bivins was also especially helpful and understanding in her feedback, and I am grateful to her for the time spent reading the work presented here, which undoubtedly helped to shape this monograph.

CHSTM was the most sociable and lively history of medicine department which I have encountered, and the staff and postgraduate community there helped to make the painstaking processes of writing and researching enjoyable. In particular, I wish to thank the PhD community who were at the centre during my time there. These were, in alphabetical order, Andrew Bowman, Joanna (Jo) Baines, Anna Carlsson-Hyslop, James Farry, Paula Habib, Emily Hankin, Valerie Harrington, Alice Nicholls, Michael Rayner, Tom Lean, William (Ray) Macauley, Marie Reindholdt, Rachael Russell, Robert Smail, Melissa Smith,

Joel Tannenbaum and Hannah Waterson. The active postdoctoral community at CHSTM were also particularly supportive. In particular, I wish to thank Michael Brown, Robert Kirk, Neil Pemberton and Duncan Wilson for their interest and support in my research.

During the development of this project, I was fortunate enough to be invited to spend time at European institutions as part of the European Science Foundation's 'Drugs History' project. I am particularly grateful to Christoph Gradmann at the Institute for General Practice and Community Medicine in Oslo and Jean Paul Gaudillère and Delphine Berdah at the Centre de Recherche Médecine, Sciences, Santé et Société, INSERM, in Paris, for discussing my research findings with me and for allowing me to present at their institutions. Hearing feedback from an international perspective was particularly enlightening.

Various presentations were given between 2006 and 2009 on aspects of this research. I wish to thank CHSTM, Northern Centre for the History of Medicine; the now defunct Wellcome Trust Centre for the History of Medicine at University College London; Society for the British History of Science; Society for the Social History of Medicine; Trinity College Dublin; University of Barcelona and the CRAASH group at the University of Cambridge, amongst others, for allowing me to participate in their events. The comments and suggestions offered by audience members were often constructive.

Following the completion of this project, I gained employment at the Centre for the History of Medicine in Ireland at University College Dublin. This gave me the opportunity and financial resources with which to complete this book. I wish to thank Mary Daly, Greta Jones and Leanne McCormick who proved to be excellent colleagues to work beside.

I am especially grateful to Catherine Cox for her patience in helping in various aspects of my career. Catherine was highly supportive when I suggested running a Wellcome Trust-funded workshop entitled 'History, Digestion and Society: New Perspectives', in April 2010 which added fresh insights to the research findings which I present here. I would like to thank the following speakers: Katherine Angel, Ann Daly, Rhodri Hayward, Erik Loomis, Rachael Russell, Funke Sangodeyi, Michael Worboys, Rebecca Wynter and Adrian Zandberg. The conference would not have been as successful were it not for the efficient organization and administration provided by Michael Liffey. Earlier chapters of this book were taught on the Masters in Social and Cultural History of Medicine group. Teaching this highly intelligent group was highly enjoyable and I thank Desmond Bates, David Durnin, Aine Mitchell, Kirsten Mulrennan and Christine Murray for their constructive comments.

The staff at the following libraries provided help and assistance: British Library; John Rylands University Library of Manchester; Royal College of Physicians of Ireland; Royal College of Surgeons in Ireland; National Archives

(UK) and the Wellcome Library. I am also grateful for the *Irish Times Health Supplement* and the organizers of the 2009 Manchester Histories Festival, John Pickstone and Mari Lowe, for allowing me to fulfil the outreach potential of aspects of this study.

I am grateful to the editors of *Medical History* and *Journal of the History of Medicine and Allied Sciences* for publishing earlier drafts of chapters which are now revised and incorporated into this book. The comments provided by the anonymous reviewers were invaluable in enhancing the project.

Finally, I wish to express my gratitude to my series editors Keir Waddington and David Cantor, who allowed me to publish in the new Studies for the Society for the Social History of Medicine series, as well as to Mark Pollard and Julie Wilson at Pickering & Chatto, who guided me patiently and efficiently through the process of publication. The insightful comments of my anonymous reviewers were also gratefully received.

On a more personal note, throughout the years, my family has been exceptionally encouraging with my career, despite feeling that I spent far too many years at university. I wish to thank Pauline Miller, Kevin Miller, Sarah Miller, Katie Miller and Miriam Trevor for their continuing support and encouragement. In particular, I wish to thank Alice Mauger for her support and patience, especially when proof-reading this monograph. Alice's intelligent feedback and encouragement were invaluable, and I wish to dedicate this book to her.

LIST OF FIGURES

INTRODUCTION: HISTORY AND THE STOMACH

This book is concerned with the stomach and its ailments within the broad period 1800 to 1945. Throughout, I will also address wider issues regarding the complex interaction between reductionism and holism which informed the understanding and management of gastric illness during the nineteenth and early twentieth centuries. In doing so, I map contested approaches adopted by different medical disciplines and investigate the nature, functions and consequences of competing discourses regarding the stomach and its physiological role. This entails an engagement with themes relating to the rise of reductionism and medical professionalism, and the function of holistic models of the body within these historical processes.

Such themes are highly relevant today as medical professionals continue to debate the role of the mind in causing stomach complaints, as exemplified by the issue of whether or not stress plays a defining role in the production of gastric ulcers. Since the discovery of *H pylori* bacteria as the most likely cause of gastric ulcer disease in 1983, its treatment has tended to become predominantly reductionist in nature, as gastroenterologists express an inclination to eliminate bacteria in the stomach.[1] This approach is contested by those wishing for greater interest in the patient and the potential role of environmental and psychosomatic factors in disease causation.[2] The extent to which the patient as a whole should be incorporated into therapeutic regimes is therefore a problematic contemporary issue.

I also argue throughout that the stomach has acted as an organ which has been imbued with high levels of bodily significance, and that there exists a long-standing tradition of emphasizing the organ's importance to medical, social and cultural life. This became increasingly clear throughout the processes of accruing research for this project, which began life as a relatively limited investigation into peptic ulcer epidemiology. What emerged unexpectedly from the primary literature was a wealth of material pertaining to the stomach and its medical complaints. Furthermore, discussion of the organ did not appear to be limited to occasional historical outbursts in activity as medical science made new discoveries relating to the workings of the digestive system, or as new disease entities became elucidated. The stomach appeared, on the contrary, to occupy a central and persistent position in the development of key fields of medicine including pathological anatomy, physiology, surgery and psychosomatic medicine. Moreover, the vast amount of popular literature produced from the nineteenth century onwards pertaining to digestion and the stomach indicated that the organ has long held corporeal significance, appealing to a wide, non-specialist audience. This cultural fascination with the organ seemed to have necessitated an engagement by the medical profession with lay concepts of what the stomach means.

What became strikingly obvious was that gastric illness has acted as an area of intense medical and popular interest during the modern period. Furthermore, there existed differing, and often competing, interpretations of how the stomach should best be conceptualized and managed at any given time. Even at the peak of reductionist medicine at the start of the twentieth century, it proved surprisingly hard to convince both the public and medical professionals that approaches to the organ which neglected the complex interaction of body and mind were entirely accurate. The complexities and debates surrounding interpretations of the stomach remain with us today. Tellingly, a quick glance at a medical or scientific dictionary reveals a tendency to conceptualize the organ in terms of its physiological characteristics alone. For instance, *Merriam-Webster's Medical Dictionary* defines the organ as:

> (1): A dilatation of the alimentary canal of a vertebrate communicating anteriorly with the esophagus and posteriorly with the duodenum
>
> (2): (a) one of the compartments of a ruminant stomach; (b) a cavity in an invertebrate animal that is analogous to a stomach; (c) the part of the body that contains the stomach (belly, abdomen).[3]

Yet *Collins English Dictionary* offers a much wider interpretation:

> 1. (Life Sciences & Allied Applications / Anatomy) (in vertebrates) the enlarged muscular saclike part of the alimentary canal in which food is stored until it has been partially digested and rendered into chyme Related adj gastric
>
> 2. (Life Sciences & Allied Applications / Zoology) the corresponding digestive organ in invertebrates
>
> 3. the abdominal region
>
> 4. desire, appetite, or inclination (I have no stomach for arguments)
>
> 5. an archaic word for temper
>
> 6. an obsolete word for pride
>
> *vb (tr; used mainly in negative constructions*
>
> 1. to tolerate; bear (I can't stomach his bragging)
>
> 2. (Life Sciences & Allied Applications / Physiology) to eat or digest (he cannot stomach oysters).[4]

What is notable about the second entry is that it implies that the organ has taken on more diverse meanings than the medical profession allows for, and which even today implies some intricate level of interaction between mind and stomach. The stomach materializes as a highly ambiguous organ, definitions of which might encompass a broad array of multifaceted cultural meanings. Ultimately, it emerges as far more than a medical artefact. In lay terminology, the 'stomach' can define the abdominal region as a whole, incorporating all of the digestive organs. For instance, the early nineteenth-century physician R. J. Culverwell stated in 1837 that, 'time was when "the stomach" represented a certain undefined ambiguity which, whether it meant the whole cavity of the trunk, or a certain void left in some particular position thereof, is uncertain'.[5] Its association with temper and pride implies that the organ was once imbued with emotional characteristics, a phenomenon not easily accounted for in strictly scientifically defined terminology which is less concerned with the interaction between the abdominal region and the brain, or the patient as a whole.

It is this historical uncertainty surrounding the stomach and its illnesses which I will explore. Historically, deeply conflicting constructs of the clinical phenomena of the organ have been presented by dieticians, surgeons, pathological anatomists, physiologists and psychiatrists, and these differing views have interacted in a complex manner. What has differentiated the stomach from other organs that came under the medical gaze such as the kidney or liver, I shall maintain, is its persistent centrality to debates taking place between contested and competing fields of medicine. The organ had a profound influence on the manner by which members of different medical disciplines attempted to convince both members of the profession as a whole and the general public of the validity of their specialist area of enquiry. Effective management of gastric illness long remained elusive, and was therefore open to high levels of professional speculation. Hence, claims to be able to diagnose and treat its myriad problems played a formative role in the development of professional relationships within medicine, and impacted upon the patient–doctor interaction. To master the stomach would have ensured wider acceptance of emergent forms of medical practice. This was of particular importance at times when disciplines including abdominal surgery and laboratory medicine were in a relatively early stage of development, and therefore searching for general acceptance as a valid diagnostic or therapeutic field.

Surprisingly, given the wealth of material available, historians have left analysis of the stomach remarkably neglected. Yet the sparse amount of available academic research masks the high levels of attention that have traditionally been directed towards the abdomen.[6] Back in 1979, Gert H. Breiger complained that medical historians were guilty of ignoring the problematic condition of dyspepsia, or indigestion, noting that the term did not appear in the index to *Bulletin of the History of Medicine* and that it was cited only infrequently in the Wellcome Institute's *Current Work in the History of Medicine*. He suggested that the historical importance of the condition had in fact been heavily underrated and that a careful study of dyspepsia held the potential to lead to a vastly enhanced understanding of nineteenth-century medical theory and practice.[7] However, little else was published on the historiography of stomach problems until eighteen years later when William F. Bynum argued a similar case in *Gastroenterology in Britain: Historical Essays* (1997), an edited volume which he compiled with the specific aim of outlining some of the area's main individuals, issues and technical developments. He anticipated that future scholars would be able to expand upon what he considered to be a highly significant, if complex, theme.[8] Within the same publication, Roy Porter claimed that 'it is odd that the early history of the disorders of the stomach and the medicine surrounding them has been so neglected'. Accordingly, he provided a historical sketch of biliousness, a broad eighteenth-century term encompassing various digestive complaints, with a similar hope that this might stimulate further research.[9]

Despite a burgeoning awareness of the potential historical significance of gastric health in the early modern period, themes relating to it in the modern era have not been expanded on in considerable depth.[10] Discussion appears sporadically, but tends to be mentioned only in passing rather than being placed centrally.[11] Overall, we are left without a firm chronology of the development of modern constructions of gastric disorders, with the exception of brief, self-congratulatory accounts written occasionally by members of the medical profession. Within these, medicine is typically presented as having marched forward towards ever more accurate forms of identification, classification and therapeutic improvement. Such approaches are inherently overly positivistic, detailing lists of discoveries, breakthroughs and technological innovations rather than critical historical analysis.[12] Cultural historians have begun to express a stronger interest in matters relating to modern aspects of diet and digestion, most notably Christopher Forth and Ana Carden-Coyne within their edited volume, *Cultures of the Abdomen* (2005).[13] However, there still exists little research on the historical nature of gastric illness itself, with the exception of culturally charged topics such as obesity and anorexia.[14]

Why, then, has the medical history of the stomach and gastric illness not been written so far? Steven Shapin has suggested, with reference to the early modern period, that this problem stems from the close association which the management of the organ has traditionally shared with dietetics, a culture relying primarily upon common sense rather than medical expertise. Hence, gastric illness has not left an obviously discernable record of ideas, innovations and techniques in comparison to other areas of medical enquiry. It therefore appears as far less dignified in comparison to those areas which do inspire narratives of scientific discovery and therapeutic innovation.[15] Yet this explanation proves problematic when applied to the modern context. For instance, peptic ulcer and duodenal ulcer disease have tended to lie more within the realm of medical or surgical treatment over the last century, as opposed to common sense dietetic management. Patients relied upon operative intervention or the manipulation of gastric acids, and later antibiotics, as their principal form of cure.[16]

A further reason undoubtedly lies in the fact that historians of medicine have often been too cautious in their willingness to engage with chronic diseases. This is problematic. The day-to-day business of medicine has always been, and still is, primarily concerned with such illnesses. As Adrian Wilson has argued, exploration of the history of chronic disease in fact takes us to the very heart of medicine's history, as it involves engaging with the ideas that determined what medical practitioners saw, and what thus helped to define what medicine actually was at any given time.[17] Yet historians have preferred to write about certain types of disease at the expense of others. Historical understandings of gastric conditions are thus far less developed in comparison to life-threatening diseases such as smallpox, plague, tuberculosis and AIDS. This argument formed a central part of Roy Porter's exploration of gout, a condition which he felt had not been treated in quite the same way as acute diseases. He suggested that this was because, unlike catastrophic mass killers, gout has not elicited elaborate rationalizations and victim-blaming tropes. Put simply, our understanding of chronic disease history is far less developed primarily because such problems are not as sensational.[18] Constipation is a prime example of this, being a condition whose past, until recently, received little analytical attention, not least because it has been perceived as a trivial complaint. It certainly proves hard to locate obvious, single major discoveries or breakthroughs relating to its management. Nonetheless, as James Whorton has expertly shown, the medical community has for centuries been forced to find ways to understand and treat the problem using a variety of ideas and methods.[19]

There are obvious differences between the management of acute and chronic illnesses, and analysis of the latter has the potential to offer a fresh perspective on understandings of the historical activity of medical professionals. Chronic problems, particularly when they arise within the gastrointestinal tract, have often proven to be debilitating, incurable and incomprehensible, particularly when located in inaccessible bodily regions that, until more recently, could not be easily or accurately visualized and accessed. Notably, conditions such as peptic and duodenal ulcer often have a long life history before they develop into more serious conditions involving perforation or haemorrhage. As well as crippling sufferers for years and causing great discomfort, distinguishing their symptoms from other forms of gastric illness baffled medical professionals for centuries.

As J. C. Eno, the famed Newcastle chemist, stated in his highly popular *The Stomach and its Trials* (1871), in diseases such as fever there is 'a speedy close, either in health, or in that final issue which puts an end to all troubles' while smallpox and measles 'come but once'. He claimed that for the victims of these illnesses 'there is some hope of a change, some prospect of a termination to their complaints, or, at least, some occasional intervals of freedom from affliction', adding that even the wasting of consumption did not deprive the patient of hope. However, there appeared to be little hope for the chronic dyspeptic. Eno suggested that the illness 'will neither kill the patient nor depart from him. Hitherto, it has been more hopeless than a sentence of imprisonment for life, for here there is a hope of a pardon.'[20] Gastric ulcer presented the medical profession with problems of a vastly different nature than epidemic outbreaks of acute medical conditions.

Furthermore, chronic abdominal illness has always existed alongside, and in the periods between, the epidemic outbreaks that have occupied the attention of medical historians. This ultimately begs the question as to which forms of medical complaints held greater significance to contemporaries living in the past? This is not to underplay the importance of epidemic diseases, as they have without doubt helped to shape many aspects of human existence and experience. Yet, as Anne Hardy has shown, between 1700 and 1830, London was completely free from major epidemics such as plague and cholera, while endemic infections such as smallpox fluctuated at a level that made no exceptional imprint on popular memory or on the Bills of Mortality.[21] An important question, therefore, is what health issues might have been considered to be of immediate significance in these lengthy periods where life-threatening diseases were less prevalent and how did their management differ from epidemic conditions?[22] Additionally, what impact did they have upon the structure and activities of the medical profession?

We should certainly not assume that just because effective management and understandings of chronic stomach conditions were late to develop in comparison to those of many epidemic diseases or medical specialities, that little of interest was happening historically in relation to them. A quick glance at the historical record might show that the British Society of Gastroenterologists was not founded until 1937, or that the journal *Gut* only emerged as late as 1960, over a century after the emergence of various other specialist forms of medical treatment. However, while the field was admittedly relatively slow to develop, this does not imply that little worthy of analytical interest existed prior to the emergence of therapeutic methods such as antibiotics or the discovery of *H pylori* bacteria's role in producing ulcer disease.[23] Medical practitioners have for centuries had to deal with stomach complaints regardless of the state of knowledge relating to them or the quality of diagnostic or investigative technologies available to them.[24] Exploration of the historiography of gastric problems therefore has the potential to open up an array of topics of interest to historians, not least because their nature differed vastly from the more commonly discussed diseases. Analysis of their past has the potential greatly to advance understandings of professional approaches to the body and the manner by which certain areas of the body can act as rhetorical sites of professional identity.

Throughout this book, I will maintain that the myriad experiences of the stomach and its illnesses, most notably dyspepsia and gastric ulcer, have persistently occupied prominent positions in modern British medicine and in the construction of professional identity. I am less concerned with the lived experience of gastric illness, and more with the medical encounter with such problems, although such a study would indeed be valuable. The patient view and gendered approaches to the stomach are both factors which require closer analytical attention than this study allows for. Furthermore, this is a fundamentally British project. As Chapter 1 will show, the stomach came to occupy a central role in nineteenth-century medico-social thinking, a response to the social changes brought about by the Industrial Revolution and the need to ensure public health. International comparison would add complexity and depth to the findings which I present here. What I do aim to achieve is a wider acknowledgement of the centrality of the stomach as a persistent marker of medical achievement and professional activity, as well as a historical index of social anxiety. I will examine the uses of gastric illness as a symbolic resource for debating medical achievement and professionality, as well as individual and national well-being in times of rapid social and economic change in Britain.

Of course, I am not suggesting that the stomach did not become conceptualized as an important entity prior to the nineteenth century. Galen (AD 129–200) affirmed the digestive processes of the stomach as an integral aspect of the maintenance of humoral balance.[25] For the ancient Greeks, the stomach ranked high in the hierarchy of the body, being denigrated as the organ closest to man's animal nature.[26] Throughout the Middle Ages, the stomach was regularly portrayed as an impure organ, devoid of rationality and spirituality, which could exert inappropriate influences upon the wider system of the body, forming a dangerous counterpart to the spiritual functions. It was believed to be hidden deep within the body due to its uncleanness and to protect the mind from its detrimental influences.[27] Notably, Van Helmont (1579–1644) rejected the idea that the soul was located in the heart or brain, positioning it instead within the cardiac end of the stomach and devoting considerable attention to interaction between digestion and bodily disorders.[28]

Yet fuller discussion of these themes lies beyond the scope of this book. My focus here is instead on the modern era. Accordingly, my chronological structure inevitably follows the shift of the medical gaze from the whole person to individualized organs and systems during the nineteenth century.[29] While, to a certain extent, this essentially follows a Foucauldian line of reductionism, I am especially interested in the tensions that emerged between the competing disciplines who were self-consciously attempting to redirect medicine.[30] Rather than tracing a straight-forward line of reductionism, I analyse the controversies that came to surround the management of the stomach within this process. Unlike, say, the introduction of the stethoscope to listen to the beats of the heart, it proved difficult to convince practitioners of the validity of new techniques designed to observe the activity of the stomach, or to treat its conditions therapeutically. One reason for this is that, in contrast to organs such as the heart, clinicians could easily cling to their traditional form of attack upon many gastric complaints if they were unsure about a new technique: dietary advice. This therapeutic regime persisted due to the inability of practices associated with reductionist medicine to convince other members of the field of the validity of their knowledge of the gastric tract. In the clinic, medico-scientific ideas regarding the organ were often adopted cautiously.

This heightened the potential of any group of medical specialists, if they did manage to gain a firm foothold in the management of gastric disorder, to enhance their medical authority in the eyes of both the medical profession and the public. For instance, as Chapter 4 will demonstrate, early twentieth-century abdominal surgeons exemplified duodenal ulcer disease as the flagship disease of their field, as it epitomized their ability to undertake research on the living patient. Yet this prompted physicians to assert their own, traditional methods of managing the problem. Similarly, laboratory medicine practitioners explored the digestive system partly in order to assert their scientific prowess throughout the period in question. Their findings too became contested by physicians and opponents of laboratory techniques, whilst they were accepted only in a limited form by surgeons. Underlying this was the persistence of dietetic approaches to the stomach, and anxiety regarding basic factors which remained obvious to the physician: the intricate connections existing between factors such as anxiety and the onset of gastric ulcers; and the inter-relationship between environmental factors and gastric illness that could not so easily be explained by reductionist medicine. What was unique about the role of the stomach within all of these contested fields was the manner by which management of its ailments was persistently held up as a potential exemplar organ of medical ability. The stomach became a central site of professional discourse.

In my discussion of the historical nature of gastric illness, I draw principally from the methodology of writers such as Charles Rosenberg. In models aligned to principles of social constructionism, disease acts as an elusive entity. It is not solely a biological event, but also acts as a construct which reflects medicine's intellectual and institutional history, legitimates public policy, acts as an aspect of social and individual identity, sanctions cultural values and structures the doctor–patient interaction. Medical thought and practice are rarely free of cultural constraint. Even medicine's more technical aspects are shaped by the demands of cultural assumptions and are moulded by the shared intellectual worlds and institutional structures of particular communities and sub-communities of scientists and physicians. Differences in speciality, institutional setting and academic training play an integral role in determining how medical professionals formulate and agree upon definitions of disease.[31] This approach holds relevance for theories regarding the 'rise and fall' of gastric ulcer between 1800 and 1950. Chronic gastric ulcer is believed by epidemiologists to have witnessed a sharp rise from the early nineteenth century onwards, and disease incidence declined only in the 1950s. Yet, as I will suggest towards the end of this study, shifts in diagnostic patterns which occur within the historical record have been analysed without fully taking into account shifting diagnostic practices and professional relationships.[32]

In my opening chapter, I suggest that in nineteenth-century Britain the stomach became widely prioritized as the most significant of the bodily organs whose dysfunction was the cause of virtually all other bodily ailments. This provided physicians with a relatively simple therapeutic regime to cure all bodily ailments: dietary advice. The heightened centrality of the stomach within what was essentially a holistic model of the body was stimulated by the rise of nervous theory and concepts of sympathetic relationships in existence between the internal organs. This facilitated the production and dissemination of an abundance of material intended to educate the general public, as well as the medical profession, on gastric illness and the necessity of maintaining a healthy stomach in order to guarantee full general health. Yet this process was not just a consequence of shifting medical constructions of the inner body. It also resulted from, and was heightened by, profound anxiety regarding modernity and industrialization, and the potential incapacity of the natural body to navigate its way successfully within a rapidly shifting British society. Hence, problems such as dyspepsia became persistently positioned as the national disease of Britain.

Chapter 2 focuses upon the emergence of more specific disease categories throughout the first half of the nineteenth century. It assesses the isolation of gastric ulcer disease from vague, encompassing diagnostic terminology such as dyspepsia, a process resulting from the rise of anatomical and pathological enquiry. It also asks why and how such a rapid shift occurred in this period from occasional observations of ulcers of the stomach towards the rapid acceptance of a systematic conceptual framework that enabled it to become generally accepted as a unique, recognizable disease entity, distinguishable from a complexity of gastric diagnoses available to contemporary practitioners. This formed an integral part of the establishment of an organ-based approach to gastric illness which provided alternative definitions and treatment than those available in contemporary literature on digestion and diet.

Controversies surrounding investigations into digestion made by late nineteenth-century laboratory scientists are assessed throughout the following chapter. The professional agenda of physiologists, I maintain, required a closer examination of the stomach than had been previously witnessed, focusing upon the organ's constituent gastric chemicals. This approach signified a marked alternative from both pathological anatomy and dietetic advice. Successfully gauging the physiology of the stomach to provide a cure held the potential to legitimize the controversial field of physiology, an area tainted by association with animal experimentation and medical cruelty. The technologies and procedures associated with the stomach became central to antivivisectionist propaganda, and later to that of the suffragettes when the stomach tube began to be employed upon them as part of the British government's policy of forcibly feeding imprisoned militants.

Chapter 4 analyses the rise of abdominal surgery around the turn of the twentieth century, and the desire of British surgeons including Berkeley Moynihan to exemplify the achievements of abdominal surgeons as a marker of surgical prowess. This was essential in an era when reductionist approaches to the body, and the stomach, were proving controversial. The stomach became the prime organ to be targeted surgically. Hence, it acquired symbolic significance in terms of surgical ability to manipulate the inner recesses of the body. Abdominal surgery also aimed to acquire a more convincing scientific rationale than laboratory medicine, whilst presenting a therapeutic regime which became contested by physicians eager to maintain the validity of traditional ways of managing the stomach.

The final chapter maintains that the Second World War was a pivotal moment in the contested discourses surrounding the management of gastric disorder. Apparently high levels of dyspepsia and gastric ulcers amongst both civilians and soldiers helped to reinforce concepts that forms of medical therapy which entailed highly reductionist approaches were proving to be ineffective, or at least problematic. Professional debates which had raged for decades culminated in calls for reductionist medicine to be reconciled with approaches stressing the importance of the mind–body relationship, and the position of the stomach within this.

1 THE NATIONAL STOMACH: INDIGESTION AND NINETEENTH-CENTURY BRITISH SOCIETY: AN OVERVIEW

In 1853, an obscure author named Sydney Whiting (d. 1875) published *Memoirs of a Stomach*, a short book which was to prove immensely popular throughout the rest of the century. It ran into various editions during the following thirty years and was even translated into French in 1888. On the surface, this might appear to be an unusual choice of reading material for such an extensive audience, given that the main protagonist is a remarkably literate stomach, named Mr Stomach, who describes the misery of his long life to the reader in painstakingly minute detail. The organ begins by detailing how his ancestry dates back to the invasion of the Saxons when the great Sir Hugh Stomach was created baron due to the huge quantities of beef that he was able to digest. Sadly, it is explained that Mr Stomach's mother died soon after giving birth to him, 'joining the stomachs of another sphere'.[1] The consequence of this was the commencement of a life of poor health, prompted by the organ's owner being breast-fed by a London woman whose milk was contaminated due to her over-indulgence in liquor and porter.

Throughout his youth, Mr Stomach complains of being forced to digest adulterated flour, sweetmeats, oysters and tobacco smoke, foodstuffs not well-suited to his delicate constitution. At college, the organ's master consumes long breakfasts lasting until noon, throughout which masses of food from around the world would be poured into his cavity. It is at this point that an attack of severe dyspepsia strikes for the first time. Large quantities of inappropriate drugs are administered by a quack physician, which worsens his already fragile health. However, recovery ensued. Shortly afterwards, the organ's hapless owner falls in love. Mr Stomach complains bitterly of his master's new-found habit of singing loudly, lamenting that he was 'constantly being woke up in the night, and found myself either walked up and down the room, the maniac repeating love ditties'.[2] The honeymoon proves to be an even more traumatic experience for the unhappy stomach as the consumption of Continental foods plays havoc with his health. Eventually, his master becomes employed in a well-paid city job which provides him with the financial resources to indulge excessively in alcohol. The final disturbance to health is caused by the organ's owner becoming involved in heated debates, which causes a wave of evil passions that disgust Mr Stomach's sensibility.

The content of this text raises numerous interesting questions. Why would such a bizarre story prove to be so popular for a Victorian audience? What causes Mr Stomach to become so alarmed by both the physical and emotional behaviour of his owner? Why is his health identified as deteriorating so rapidly in the specific context of urban life? What concerns facilitated the mapping of such intricate connections between an organ hidden deep within the body and society at large? I shall suggest that Mr Stomach's memoirs are illustrative of the organ's pivotal positioning in constructs of both the healthy and unhealthy nineteenth-century bodily system. Broadly speaking, until this period, the stomach had not been prioritized in the same way that, for instance, the heart had been following William Harvey's publication on blood circulation.[3] Yet many medical men began to imbue the organ with especial significance. Abdominal complaints became identified as problematic on two encompassing levels: individually and nationally. As Whiting's *Memoirs of a Stomach* reveals, there appeared to have existed intricate connections between the stomach and certain aspects of British society. Dyspepsia is presented in the text not just as a medical diagnosis but also essentially as a behavioural and moral problem. This chapter aims to unpack this complex interaction between individual and collective gastric health.

The Nervous Stomach

During the late eighteenth century, new conceptualizations of the workings of the inner body invested the stomach with enhanced significance. This is not to suggest that the abdomen had not been identified as a prime bodily region in earlier periods. Matters relating to diet, nutrition and digestion had certainly been taken seriously at least since ancient times.[4] Yet a growing acceptance of models of the human body that stressed the relevance of the nervous system precipitated shifts in the way in which the interrelationship between different organs was conceptualized and managed.[5] The most influential figure in this process was undeniably William Cullen (1710–90), a prominent Scottish doctor and chemist who viewed all aspects of neurophysical life as stemming from nervous energy. Cullen developed an elaborate taxonomical system that effectively reordered thinking about disease by allocating a prioritized role to the nerves.[6] His nosology was published in 1769, and was to be frequently republished and translated well into the nineteenth century. The most significant aspect of Cullen's work was to be its establishment of the use of the term 'neurotic', a diagnostic phrase that became an essential facet of medical theory and practice throughout the subsequent century.[7]

The relevance of the stomach's bodily role was inadvertently enhanced by an enthusiastic uptake of principles relating to the nerves. The Edinburgh physician Robert Whytt (1714–66) was a central figure in this process, having been a key character for late eighteenth-century neurology. Crucially, he developed notions of what he termed 'sympathy'. This term was employed to describe how different parts of the body were seen to interact with one another. It depicted a system rich with links between organs, all connected by the nervous system. Notions of nervous sympathy provided an adaptable mechanism for explaining how pain or discomfort might be felt in organs far away from the initial seat of disease. A body whose constituent parts were fully connected by an intricate network of nerves might result in body parts becoming too much aligned in sympathy with one another. This was seen to encourage a misleading catalogue of diagnostic symptoms. Whytt's model therefore elucidated the problem of defining how the symptoms of gastric problems might be confusingly felt in the heart or lungs, a phenomenon which added bewildering complexity to the categorization of disease and illness at the clinical level, and which ultimately seemed to hinder the physician's diagnostic capabilities.[8]

Concepts of sympathetic affection were not unheard of in earlier periods. Since ancient times a wide range of medical writers had directed attention towards the complex interactions seen to exist between the variegated organs of the body. The earliest known written accounts of sympathy can be found in the work of Galen.[9] The influence of Greek ideas persisted throughout the centuries although they eventually became intertwined with ideas relating to occult and cosmic forces by Renaissance physicians.[10] It was Whytt, however, who popularized the theoretical construct that a primary function of the nervous system was to maintain the natural harmony of the body parts. He identified two different forms of sympathy: a general one extending through the body, and a specific one in existence between its major organs.[11] Yet nervous energy was not diffused evenly throughout the system. Certain areas contained more of it than others. Hence, there emerged potential to identify bodily locations which might act as a locus of nervousness, and which held greater importance in terms of their potential sympathetic consequences.

Attempts were made to enhance the role of the liver. Yet these became heavily critiqued primarily as, upon anatomical examination, the organ was observed to have discernibly fewer nerves, dull sensibility and hardly any sympathetic connections with other organs.[12] The uterus, too, had long been viewed as a source of female emotional and behavioural anomalies, a concept with practical implications which persisted well into the nineteenth century. This, however, provided a model applicable only to the female sex.[13] For a substantial number of authors, the stomach became a prioritized site within the human nervous structure. The organ became persistently delineated as a mechanism containing an abundant supply of nervous energy and which held the important function of dispensing this throughout the body. Such models proved remarkably durable throughout the nineteenth century. Accordingly, the stomach can be found throughout the period invariably described as 'a focus of vitality – the centre of a department in which the living principle is most abundant and exquisite', the 'foundation or root of the complex apparatus', the 'great nervous centre or sensorium of organic life' and even as the 'great abdominal brain'.[14]

Contemporaneously, strands of biological theory began to stipulate that the digestive processes were the central facet of all organic life. One popular theory claimed that as the lowest forms of life had no sense, pulse or motion, they were essentially animated stomachs. Digestion seemed to be their sole faculty. The sponge provided an illustrative example of this, a creature that appeared to consist almost entirely of minute pores, described by one author as 'many little mouths, which perpetually suck in the sea-water, and the animalcules with which it abounds'.[15] The sponge was one of the most primitive forms of life, and it appeared to do little else with its life but digest. It was, essentially, a swimming stomach. Conclusions were developed drawn upon this theoretical basis that the stomach was the most basic, and therefore the most important, of all the organs.

Elevating the stomach into an organ of key significance became a central aim within many circles of medical thinking during the opening decades of the nineteenth century. The most extreme proponent of this view was John Abernethy (1764–1831).[16] Abernethy was a highly influential anatomy teacher, a famous surgeon at St Bartholomew's Hospital and the founder of its medical college. He professed himself to be the spokesman for the physiological and pathological views of John Hunter (1728–93) who, like Whytt, had placed especial emphasis upon the inter-relation of the organs. Hunter portrayed bodily parts including the teeth, mouth, throat and stomach as interacting components of a physiological system of food ingestion and digestion that was characteristic in all animal life.[17] Somewhat surprisingly, given Abernethy's overt allegiance to Hunterian principles, Hunter had relatively little to say on the subject of nervous sympathy within his written work.[18] However, early nineteenth-century London surgeons picked up upon, and grossly exaggerated, his sparsely illustrated concepts of gastric sympathy. Abernethy in particular took Whytt's views on sympathy to a fanatical level, campaigning tirelessly for wider recognition of the bodily significance of the stomach, its illnesses and the distressing consequences of gastric sympathy.[19]

Abernethy's books proved highly popular, not least because he spoke of them obsessively, showed them in every one of his lectures, wrote to *The Times* about the theoretical frameworks contained within them and even dispensed free copies to his many patients. Tellingly, his *Surgical Observations on the Constitutional Origin and Treatment of Local Diseases* (1811) ran into eleven editions and became known by others as 'My Book', a sarcastic dig at his incessant conversational references to it.[20] He also adapted his principles for a general audience in the popular short piece *The Abernethian Code of Health and Longevity* (1829).[21] His influence was such that, later on in the century, the *Observer* claimed that Abernethy's development of concepts that all bodily disease was traceable to gastric derangement was one of the greatest services ever rendered to mankind. The author went so far as to assert that it was one of the most valuable contributions to medical literature made since William's Harvey's discovery of the circulation of the blood.[22]

Like many of his contemporaries, Abernethy noted the seemingly central role of the stomach to general animal life. For him, the biological make-up of the most basic of animals appeared to be so simple that many of these creatures could be dissected whilst alive, but would not die. Instead, each part transformed into a living entity in itself. This was characteristic of the aforementioned theory that such basic creatures did nothing but digest. For Abernethy, they essentially consisted of one organ whose sole function was to assimilate food. This digestive organ retained its pivotal function as complex life developed, although its workings became more complicated as increasingly intricate interactions with the other parts evolved. More advanced forms of life, including the human, were rendered distinguishable from the basic organisms due to this complex interaction, the purpose of which was to maintain the well-being of the system as a whole. Abernethy ultimately provided a model of the system that was intrinsically circular in nature, with all parts helping to modify and improve the general biological and physiological well-being of the human framework.

Abernethy prioritized the stomach above all organs within this integrated system. It is no exaggeration to claim that he was obsessed with the stomach. For him, the simpler animals could be seen to survive without nerves, lungs, blood vessels, a heart or brain, yet he found not one without a stomach. Placed at the core of the human system, he depicted it as the regulator of the motions and sensations of the entire system; the centre of power and motion from which the vital principle flowed into the different parts; the seat of muscular exertion and fatigue; and the receptacle of food, poisons and medicines, all of which would be propagated throughout the body via the complex workings of the nervous system.

Yet by prioritizing the role of the *healthy* stomach in maintaining the well-being of the general system, Abernethy helped to raise levels of anxiety over the variety of ailments that might potentially arise from an *unhealthy* stomach. A metaphor which he frequently invoked was the comparison of the role of the organ to that of the kitchen in the house. If this room was not in full working order, he explained, then every room in the house would eventually become affected. As blood seemed to be transmitted around the body following its conversion from food within the abdomen, the functions of the digestive organs needed to be strong and healthy in order to prepare and assimilate nourishment for the whole body.[23]

Despite this interest in blood and the circulatory system, Abernethy remained especially concerned with nervous sympathy. Phenomena such as the impact of a blow to the stomach producing disordered mental sensations, or mental conditions such as worry reducing the appetite, seemed to him to take place independently from the circulatory system. He found explanation for them instead within contemporary models that stressed the internal significance of nervous associations. Given this prioritization of the stomach, it is unsurprising to find that Abernethy visualized the best means of attack on such problems as being through the digestive tract. By strengthening the stomach, he maintained, the nervous ailments of the entire body might be cured or, at the very least, eased. Furthermore, this was easily achievable since diet was readily adjustable.[24]

How representative were Abernethy's views of those of the medical profession at large? Few physicians seemed to have become quite as worked up with excitement about the stomach as Abernethy. Yet a broad consensus emerged that the organ held physiological significance, and the prioritization of the stomach's role became a fashionable and pervasive theme in British medicine. Tellingly, the medical obsession with the organ reached such a peak that in 1838 the *Dublin Journal of Medical Science* complained that, for the medical community:

> The stomach is everything; heart, brain, nerves, &c., are merely secondary, and are besides entirely governed by the stomach, the ruling centre of the outward as well as the inner man. Physiologists too, when writing on digestion, and anatomists on the digestive apparatus, are apparently under the dominion of this fixed idea, this monomania.

The reviewer went on to warn that modern physiology was rendering more noble organs and functions less important, 'deposing the mind and brain from their high place, and substituting the stomach in their stead'.[25]

The influence of notions of gastric sympathy can be found being applied throughout the entire spectrum of the nineteenth-century medical hierarchy. Notably, in the 1850s, Dr Lamb, an Exeter physician, made a living by proposing that his curative methods were superior to most on offer because of his emphasis upon the abdomen. He claimed, for instance, to be able to relieve the symptoms of tuberculosis via the correction of digestive faults. For him, a diseased state of the lungs was undoubtedly a consequence of the sympathetic interaction existing between the stomach and other internal body parts. Lamb utilized methods that invigorated the stomach and constitution, thereby helping permanently to throw off the progress of tuberculosis.[26] He also claimed in his advertisements to be able to cure scarlet fever by similar means. After one patient had consulted numerous doctors, and had become increasingly weak, Lamb identified the stomach as the seat of the disease. By restoring this organ to health, he claimed to have made cardiac pain disappear, while breathing was reported to have become strong and clear. The patient was soon able to return to his occupation as a joiner and builder.[27]

More prominent medical figures also began to incorporate the sympathetic role of the stomach into their clinical recommendations. John Howship (1781–1841), a prominent surgeon, stressed that clinicians should incorporate abdominal symptoms into their investigations of violent palpitation of the heart.[28] Likewise, James Johnson (1747–1845), famed editor of the *Medico-Chirurgical Review*, suggested that irregularity of the heart was a common consequence of stomach disorder, and warned that this was regularly unsuspected by practitioners and patients. He claimed to have observed various cases initially diagnosed as chronic indigestion but which had later turned out to be masking life-threatening varieties of heart disease, a predicament that encouraged many physicians to examine the other organs minutely and to ignore the symptoms of the one where the complaint was actually located.[29]

A further illustrative example is provided by the death of the eminent sculptor Sir Francis Leggatt Chantrey (b. 1781), who died at home in November 1841 following a busy day erecting statues in Norwich. His ante-mortem symptoms initially appeared to focus around his heart. Despite being cheerful all that evening, he suddenly fell ill, sank down in his chair and died with a tin bottle of hot water pressed to the organ. Yet his physician, Dr Richard Bright (1789–1858), stated that Chantrey had suffered for months from severe indigestion, with a symptom of this being blood in the heart. Although Bright concluded that he had died from a spasm of the heart, he thought that this in itself was a consequence of a stomach complaint.[30]

The heart was not the only organ considered to be closely connected to the gastric organs. The perceived interaction between the stomach and brain, for instance, appears to have become of special interest.[31] The conditions considered liable to arise from the seemingly close relationship between stomach and brain were varied. Perhaps one of the least worrying of these was the 'dyspeptic headache', habitually dismissed as temporarily resulting from the consumption of unsuitable food.[32] Seasickness provided a further example of sympathetic phenomena. It seemed that this could cause the total suspension of the digestive powers, with the patient developing into little more than a skeleton whilst at sea. Yet upon reaching shore the brain seemed quickly relieved from distress and the digestive powers would rapidly restore.[33]

Sensory dysfunction was also regularly attributed to gastric sympathy, with dimness or cloudiness of sight being regularly perceived as a direct result of a disordered stomach. Hallucinations were also portrayed as stemming from gastric sympathy. Some of these might be relatively minor such as the appearance of black specks. Yet if it involved the misperception of large physical objects then this was considered to be indicative of a far more serious complaint.[34] Prolonged cases of gastric problems were also observed to lead to more persistent and problematic mental states. Hypochondria was regularly perceived as a particularly critical outcome of the nervous sympathy between stomach and brain, leading the patient to suffer from morbid mental diseases, and to become tormented by imaginary sense of pains or disease and turn wearisome or apprehensive of life.[35] One anonymous author argued that

> the imaginary evils thus induced, are more painful to be borne than the most acute bodily suffering; the mental misery commences with anxiety, and rapidly runs through all the grades of a disturbed mind, until it ends in melancholy, despair and perhaps suicide.[36]

Even when elaborate theories identifying the role of germs in disease causation had been developed, late nineteenth-century writers continued to emphasize the necessity of maintaining strong digestive health. Accordingly, as late as 1886, the esteemed St Bartholomew's physician Thomas Lauder Brunton (1844–1916) wrote of girls in whom attacks of consumption had been brought on by an unfortunate love affair. Although accepting the view that consumption was directly linked with the tubercle bacillus, he noted in one patient the depressing effect of a romantic disappointment in weakening the digestive strength. It had impaired nutrition and rendered the body more liable to act as a suitable locus of bacteriological infection.[37] Examples such as these are representative of the manner by which many nineteenth-century physicians approached the stomach and evoked awareness of its bodily role.

The stomach clearly maintained a pervasive role in the nineteenth-century medical imagination. Theories expounding its relevance were accepted and practically employed by men throughout the medical hierarchy and proved to be readily adaptable to changes in medical theory and practice. They also offered a relatively straightforward form of therapy for virtually all problems that the physician might encounter: simple measures of dietary improvement. It is perhaps plausible to imagine that the popularity of such theories rested to a certain extent in the ability to provide therapeutic reassurance where medical knowledge was lacking or deficient, and equipped the medical profession with ideas relating to preventative medicine that could inspire confidence in the profession. Yet this is not to say that such methods were altogether therapeutically barren. Theories of nervousness and gastric sympathy, in this instance, proved to be translatable into a common sense model of health and illness that stipulated that healthy eating and care of the stomach would result in a healthier bodily system.

The Stomach and British Society

Medical consensus on the stomach's importance was not achieved as a result of shifting medical and biological views alone. The stomach became prioritized also because its ailments became emblematic of wider social and cultural concerns. Within virtually all nineteenth-century literature on digestion, gastric complaints are framed as potentially problematic not only for the unfortunate individual crippled by abdominal pains, but also to British national health and progression.[38] Advice given on digestion rarely dissociated these themes.[39] Linkage drawn between nationality and individuality was the product of observations that increasing numbers of people in Britain were being subjected to external influences which were having a detrimental impact upon their nervous system. As the stomach was seen by many medical authorities to be the primary organ of this system, it persistently became the first bodily area to be targeted.

The nineteenth-century medical profession frequently warned that stomach problems were on the rise in Britain. In 1826, the *Medico-Chirurgical Review* stated that 'there is no complaint more common in this country than an imperfect condition of the stomach'.[40] Twelve years later, the *Dublin Journal of Medical Science* stressed that 'stomach diseases are of every day occurrence; they form the national malady of Britain, and consequently the prime staple of the medical art'.[41] Throughout the early 1850s, advertisements for Jones' Tremadoes Pills suggested that indigestion was the 'prevailing evil of the human frame, and the fashionable disease of the age'.[42] As late as 1886, adverts for Seigel's Syrup declared that 'the national disease of this country is indigestion'.[43] Tellingly, the London family physician Aldolphus E. Bridger opened his *Demon of Dyspepsia* (1888) by claiming that ten years of medical practice devoted largely to this branch of medicine had convinced him that the subject was one not only of the greatest individual significance, but also of the highest national importance.[44] The so-called 'demons of dyspepsia' were also immortalized in etchings by George Cruikshank (see Figure 1.1). Gastric illness was presented as far more than just a physical concern for the individual to conquer. It posed a communal threat.

These concepts also permeated non-medical literature. An article published in *Blackwood's Edinburgh Review* in 1861 dramatized the predicament to such an extent that it was claimed that not only was England the country most liable to gastric conditions, but also that whilst labouring under such attacks, 'nothing but family considerations prevented him [the Englishman] from blowing out his brains with a pistol, or effectually ridding himself of his woes by plunging

Figure 1.1: 'Indigestion, A. Growquill Esq.'; etched by George Cruikshank (London: S. Knight, 1835). Image courtesy of the National Library of Medicine.

into the muddy torrent of the Thames'. The author went so far as to speculate that only a fraction of the dyspeptic British had the courage to abstain from self-destruction during the gloomy months of November and December, a period when multitudes of corpses of sufferers from crippling gastric diseases would supposedly be swept across the nation's rivers.[45] Another melodramatic assessment was provided by Bridger in 1888, who suggested that:

> The fate of individuals, the interests of society, the lots of empires, the peace of families, are alike held captive by the capricious tyrant (dyspepsia) that tramples down without compunction, law, morality and religion; all which, let me repeat, hinges upon the results brought about by this extraordinary complaint, in its searching upon mind and body.[46]

Inevitably, some scepticism emerged regarding the extent to which this alarm was real or imagined. In 1821, an article in the *Edinburgh Medical and Surgical Journal* was highly critical of the large numbers of publications being produced on digestive disorders, not least because they appeared to be doing little to advance medical knowledge, acting more as a response to fashion rather than to an actual rise in stomach problems.[47] Similarly, the London physician David Uwins (1780–1837) sardonically stated that if the English continued to be dyspeptic, physicians had the potential to amass fortunes by selling advice.[48] A further medical journalist dismissed the 'disorder of the digestive organs of John Bull' as a 'delusion' and a 'fashionable folly'.[49] Late in the century, in 1884, Thomas Clifford Allbutt (1836–1925), a physician with a self-proclaimed reputation as a 'stomach-doctor', also observed that he had been surprised to find that he had treated only a handful of dyspeptic patients whilst compiling his annual statistics. He stated:

> 'How could this be?' I asked myself repeatedly. 'Martyrs to dyspepsia' are to be found at every street corner, and are said to form something little less than the staple of those who drift from consultant to consultant ... as I turned to the journals of the day I read of men of lofty endowments whose lives became accursed by dyspepsia ... our medical journals tell us the same story. Every large drug-house has its pepsins, its dinner-pills, its cordial bitters, testifying not only to the general, but to the medical cry for help against the demon of dyspepsia.

Allbutt concluded by asking 'how are we to explain this catholic wailing over a disease which is not? This wealth of balsams for sufferings which cannot be named?'[50] Yet regardless of the reality of the situation, the perceived pervasiveness of stomach complaints in Britain seems to have been taken seriously by a significant proportion of medical authorities. Furthermore, themes relating to a national digestive problem proved remarkably durable, which is very suggestive that there existed a highly receptive audience.

The postulated causative factors of an apparent prevalence of stomach complaints drew upon a complex interaction of social and biological interpretations. Linkage was drawn between the inner body and society in various aspects of medical theory and practice as the health of the nervous system was seen as being deeply intertwined with wider social or cultural concerns.[51] This facilitated the production of a rhetoric surrounding gastric illness that encompassed discussion of both the physiological and social characteristics of nervous illness. The work of James Johnson provides an excellent example of the way in which the interaction between nerves, the stomach and society was portrayed. He penned his *Essay on Morbid Sensibility of the Stomach and Bowels* in 1827. Within this text, particular emphasis was placed upon the dual role of 'nervous sensibility'.[52] Johnson claimed that the stomach's reaction to the stimulus of food was just as 'sensible' or sensitive as the eye was to light. When the organ was in a sound state of health, the sensation of food being digested would be unconscious. No sensations would be felt. Yet the substances ingested into the organ, whether dietetic or medicinal, would produce responses in the nerves, which Johnson described as 'excitement'.[53] Again, in a state of full health, this excitement would not be observed by the owner of the organ.

Yet imperfect digestion distorted this unconscious process. If too much of any substance was introduced into the stomach, 'sensible excitement' would result, which would lead the individual to feel a sensation of something happening within him. This was defined as irritation or 'morbid sensibility', a condition with the potential to cause pain, disordered bodily functions and a variety of sympathetic disorders and morbid sensations that tormented often distant parts of the body or the mind. Johnson suggested that any awareness of sensation stemming from the stomach was nature's way of informing the owner that one of its dietary laws had been broken. As he eloquently phrased it, 'to feel that we have a stomach at all is no good sign'.[54]

Johnson's interpretation of internal, digestive physiology was simultaneously coloured by a range of social and cultural concerns. In fact, he claimed that one of his main aims whilst writing *An Essay on Morbid Sensibility of the Stomach and Bowels* had been to suggest that there must exist some connecting link or prevailing error common to all classes of British society that was rendering the nation's inhabitants especially susceptible to gastric sensibility. For him, the problem of communal indigestion was becoming so severe that:

> It knocks at the door of every gradation of society, from the monarch, in his splendid palace, down to the squalid inhabitant of St. Giles or Saffron Hill, whose exterior exhales the effluvium of filth, and interior, that of inebriating potations. No moral attributes, no extent of power, no amount of wealth, are proofs against this wide-spreading evil. The philosopher, the divine, the general, the judge, the merchant, the miser, and the spendthrift, are all, and in no very unequal degree, a prey to the Proteian enemy.[55]

Accordingly, he offered his work as one not only of potential advantage to the medical profession and its patients, but also to British society at large. His work therefore held collective as well as individual possibilities for medical intervention. By doing so, he also identified the root of the problem as lying in some negative aspect of nineteenth-century British life.

Ascribing linkage between a country and particular medical problems was not a new practice. For centuries, medical thinking had assumed that the individual was born in the best, and most healthy, place for him or her to inhabit, where there was plentiful access to an environment and diet best suited for his or her constitutional make-up. Put simply, England was the most suitable environment for the English as the constitution of the country's inhabitants was sharply influenced by the nature of their place of birth.[56] These assumptions were maintained throughout the nineteenth century. Tellingly, the effects of the peculiar constitution of the British environment upon the stomach seemed strikingly observable when visitors from abroad entered the country and almost immediately began to suffer from crippling gastric problems. When the Indian nationalist Ram Mohan Roy (1772–1833), described as 'the first Hindu of any consequence' by *The Times*, arrived in Britain in 1833, he soon began to suffer from agonizing stomach problems. This was thought to have contributed to his sudden and untimely death in Bristol. It was reported that during his residence in England he had never had good health due to being 'oppressed by the climate' and that his bilious attacks resulted from a 'disordered state of his stomach caused by Britain's dense atmosphere'.[57] An environment that was naturally good for the British was therefore not necessarily seen as beneficial to those accustomed to their own national environmental constitution.

Yet this peculiar environmental make-up of a country was also seen to render its inhabitants liable to certain diseases. It is worth noting that the term 'environment' itself typically encompassed a wide range of physical, biological, hereditary, social and psychological conditions of life.[58] Weather, for instance, was only one of a variety of potential predisposing causes of general stomach disorder.[59] Some medical authors argued that the influence of the British spring acted as an exciting cause in those predisposed to gastric complaints. Given the obvious relationship between the abdomen and diet, it is unsurprising that food became a central topic in discussion of British gastric illness. Contemporary medical theory claimed, for instance, that in cold climates the organ appeared to prefer meat and ardent spirits whilst in warmer regions man seemed to subsist primarily on bread and fruits, preferring wine to strong brewed and distilled liquors. Accordingly, the French were considered to be far more sober than their German neighbours because their country's warmer temperature meant that they drank more wine and less ale and rye-brandy. The Spanish, meanwhile, were thought to be a more sober race than the French as the country's citrus fruits were more 'perfect' than in France, rendering wine and stimulant spirits less necessary.[60]

Yet national dietary choices could go wrong if left unregulated. Ireland's Great Famine (1845–51) seemed to exemplify this point. Throughout the early nineteenth century, medical writers frequently alluded to the different physiological make-up of the Irish and English races. For centuries, the Irish had famously lived on a diet consisting primarily of potatoes. British observers dismissed this as sign of inherent national laziness, as the crop was easy to cultivate. The popularity of the potato seemed to lie in the low maintenance of its cultivation, which allowed the Irish to spend more time indulging in their alleged favourite activities: drinking too much, procreating and reproducing.[61]

Of course, these claims need to be taken with a sizable pinch of salt. Discussion of the Irish by British authorities was often clouded by racial prejudice. At worst, the Irish were viewed as inhuman savages likely to contaminate the wider British social body if their errant behaviour was not held in check.[62] Yet contemporary discussion of the Irish diet is revealing about the popularity of concepts that different races might hold different internal physiology. In 1846, *The Times* claimed that the physiology of the Irish stomach had been analysed by experts and that the dependence on potato had definitely led to the communal enlargement of the stomachs of the entire country's peasantry, who now collectively craved excessively large quantities of food. Drawing upon apparent 'physiological facts accrued from dissection', one journalist suggested that a potato-fed peasant typically possessed 'a stomach of nearly twice the ordinary size' of that of the London coal-whipper, who,

> living on bread and beef ... working and sweating like a horse all day, and from morning till night going at a jog trot on a plank with half a hundred weight of coals on his back, ever consumes, or can consume, more than 4lbs or 5lbs weight of food per day.[63]

The deficiency of nutriment (inaccurately) believed to be inherent in the potato diet was thought to have been long compensated for by an increased bulk of food. As a result, the stomach of the average Irish citizen had undergone an unnatural enlargement from which it needed to recover now that it was obvious that a potato diet was not sustainable.[64] These principles were adapted for practical use. Evidence taken before the Poor Law Commissioners during the Famine stated that quantities of up to fourteen pounds of potatoes might be consumed by an Irish peasant each day. When the Irish pauper was first admitted into the workhouse, and had breakfasted on half a pound of oatmeal, so the author claimed, he appeared to be only half satisfied as his stomach wanted, as one contemporary phrased it, 'stuffing out with quantity'.[65]

The complex story of medical intervention in Irish dietary requirements in both the Famine and post-Famine periods is a fascinating topic yet to be fully investigated. Yet its importance here is the incorporation of the rhetoric of a 'national stomach' that was distinguishable from that of other countries due to racial differentiation. Of course, this involved putting together an ideal prototype of the healthy British person which failed to take physiological variation into account. This was typically the robust, healthy, sober and temperate person, perceived as being best able to contribute effectively to society. Other bodily types were portrayed as having strayed away from this ideal, holding physical and mental attributes perceived as detrimental to national progress and ascendancy, and which required the most correction.[66]

The theoretical construct of the 'national body' was a highly adaptable concept that could be employed in a variety of contexts.[67] Criticism of Corn Law taxation in the *Manchester Guardian* in 1844 stated that the present system was 'a tax not merely upon the head, but also upon the feet, upon the body, upon the stomach, upon the hands'.[68] Another publication, writing on the subject of poverty four years later, suggested that 'a revolution of the stomach is inevitable, provided there be not a change in our social system – hunger is a hard taskmaster; men, women and children must be fed'.[69] In the 1850s, the British government was described as a 'cancer in the national stomach', in a critique of its alleged excessive consumption of national resources.[70] Meanwhile, in 1858 *Lloyd's Weekly Newspaper* complained that revolution might one day occur and would arise from the empty stomachs of the nation. The author claimed that 'the people act when goaded by distress – never when strong in reason. They are polypi in the social scale; mere stomachs – sensible only when the functions of the stomach are disturbed.'[71]

Even when the disordered stomach was discussed in terms of the individual, fears were raised that resultant sympathetic effects on the mind and personality might have disastrous national consequences should the victim happen to be a political or intellectual leader. Tellingly, John E. Clarke, physician to the London Homeopathic Hospital, boldly claimed in 1888 that:

> If Carlyle had not ruined his digestion by excessive indulgence in tobacco, his influence on the world would have been happier, and he would never have written half the bitter things he did of all his friends and acquaintances; and if Darwin's stomach had recovered from the effects of seasickness, he would doubtless have been a happier man, and his view of humanity might possibly have been a more generous and exalted one. I have no manner of doubt that the writings of some pessimistic philosophers, which modern would-be thinkers waste their energies in trying to understand, are the pure products of disordered digestion.[72]

Historicizing the Stomach

How had the British nation managed to get itself into such an unfortunate predicament of being plagued by gastric complaints? The answer to this seemed to lie in the historical processes firmly embedded within medical theories of nervousness. Organic models of the nervous system provided allegoric images of national bodies that, just like the human body, never remained static. They grew and developed in line with the individual growth of its many component parts. Yet, also like the human body, parts of this system risked becoming subject to illness and disorder, meaning that reorientation towards healthier directions was required if a full state of vigour was to be maintained.

In the nineteenth century, the British 'national body' was in a remarkable state of flux. Economic, social and urban behaviour was perceived to be becoming more and more nervous as the stresses of modern life intensified.[73] As George S. Rousseau has claimed, nerves became the barometers of modern life. As civilization advanced, its inhabitants faced an increasingly greater risk of becoming subject to varieties of nervous illness. Alarmingly, between the eighteenth and nineteenth centuries, the implications for nervous problems to strike all social classes increased.[74] A particularly influential author on this theme was the Edinburgh medical philosopher Robert Verity. In 1837, Verity hypothesized that alterations to man's internal nervous structures had occurred naturally over time contemporaneously with changes in society.[75] He deemed savage man to have possessed a grossly underdeveloped nervous system and small brain, and proceeded to map the historical development of the body to the progression of mankind towards advanced forms of civilization. This process facilitated the development of extra neural pathways, exquisitely developed fibres and muscles and increased brain size. This natural process of adaptation was conceived as a necessity that allowed man to function fully whilst residing under the influences of increasingly civilized life.

For Verity, this analysis of historical aspects of the nervous make-up of human society would, provided it was interpreted correctly, reveal corresponding physiological states and relationships that held vital relevance to contemporary clinical practice. He aimed to provide a study that could be employed by physiologists who could analyse and predict how factors such as formative nutrition had, for centuries, been silently modifying the interior of the body and enlarging the boundaries of the primitive nervous structures. Changes might be foreseen, and clinical practice, it was hoped, could be orientated towards predicting and managing acute and chronic affections. Historical analysis of the nervous system provided a reference point with which to understand better the internal organization of modern individuals and the social causes of illness.

Yet Verity also intended his findings to act as a warning to Britain. This was a natural process with highly destructive national potential. He believed it to be essential that clinical practice began to pay closer attention to the historicity of the relationship between society and nervousness. This necessitated the behaviour of practitioners to be modified in order to help patients sustain nervous energy in proportion to the additional degrees of delicacy and refinement brought about by an advancing society. If not, nervousness risked emasculating the character of civilized nations, rendering entire countries vulnerable to breaking down under its pressure. What alarmed Verity the most was that he envisioned himself as residing at a historical turning point where the previously gradual process of civilization was accelerating. The nervous structure of man's natural body, however, had not been able to adapt at this quickening pace. The national mass of stomachs and nervous systems had yet to adjust to the problematic requirements imposed by the new conditions of modernity. Worse still, this dilemma seemed especially pronounced in Britain, a nation experiencing an unusually rapid march towards urbanity, industrialization and modernity.[76]

Linkage between urbanity and the stomach was far from new. During the eighteenth century, urban, consumerist trends had been blamed for encouraging negative moral influences, most often gluttony, that increased communal levels of chronic gastric complaints.[77] The Edinburgh physician George Cheyne (1671–1743) had launched the conceit that the English were not only predestined to suffer from nervous ailments but were in fact privileged to do so. Suffering from a stomach problem was postulated to be an inevitable consequence of the wealth and success of the nation, a development worth the drawbacks in abdominal health. Within such depictions, urban abdominal disease acted as a positive emblem of an economically vigourous nation. It was a condition to be proud of, a shared symbol that the nation was able to eat well.[78] If we consider the fact that the eighteenth century was a period in which most European countries were ravaged by the effects of famines and food shortages, this was not an insignificant claim.[79]

Yet Cheyne had not anticipated the rapid extent to which urban life was to become such a dominant characteristic of British life. If he had been able to foresee this, his views on the positive symbolic value of British abdominal disorder might have taken on a very different shape.[80] Between 1820 and 1914, British economy and society became more extensively and intensively urbanized than ever before.[81] Not only was the rate at which people became concentrated in large, dense and complex settlements unparalleled, but the relationship of these new urban areas to British society as a whole was shifting.[82] This was linked with a parallel process of rapid industrialization, which involved an increase of the intensity of labour and technological change.[83] Gastric health continued to be perceived as a fundamental element of urban living and modernity. Yet its positive value was difficult to apply to the context of nineteenth-century Britain. As I shall expand upon below, the predisposing factors surrounding the urban individual were seen to be becoming more varied in town and city life and to be affecting all social classes equally. Furthermore, urban indigestion held the potential to afflict constantly rising numbers of British people as emigrants from the countryside descended upon the towns and cities in vast numbers.[84] The body of the British citizen resided within unnatural, accelerated times, yet was unable to cope with its new demands. The modern urban individual needed somehow to ascertain the means with which to maintain the health of his body in accordance with nature's requirements, even if he or she happened to be situated within a variety of scenarios to which the body's nervous structure had yet to adapt.

Contemporary literature on digestion explicitly envisioned the urban environment as an artificial obstacle which the natural stomach had to negotiate its way through in order to ensure full bodily health. The countryside, it had long been claimed, provided a much healthier environment than the town or city. There, health was encouraged by fresh country air, hard labour, meagre diet and a lack of luxury. Rural areas were romantically constructed as a healthy norm whilst urban living acted as its antithesis: an unnatural departure that threatened to incur numerous health risks for its residents.[85] In particular, the city brought the patient face-to-face with a vast array of predisposing causes with the potential to trigger an irreversible decline in gastric health, a process likely to commence with illness of the stomach and which threatened to spread throughout the system via the complex entity of nervous sympathy.

Life in the nineteenth-century town appeared hazardous for the digestive tract for a variety of reasons. For James Johnson, the British had always lived in a highly changeable climate, a factor which had long held the potential to produce poor abdominal health. He suggested, however, that constant atmospheric change had little effect on the health of those living in the pure, open air of the countryside. Yet in cities and large towns, air had become imbued with millions of miasmas, the effects of which, once inhaled, were a puny or capricious appetite, and imperfect digestion.[86] Alexander Philip Wilson Philip (1770–*c.* 1851) attributed a greater numbers of hearths in condensed areas as a primary causative factor of gastric illness as it resulted in dramatic differences between the temperature of the countryside and the metropolis. The latter thus became subject to high levels of dampness, a factor which debilitated the nerves.[87] Concerns such as these proved persistent throughout the century. The prominent English homeopath John Henry Clarke (1853–1931), for instance, suggested that the 'daily bottle-of-port man' could not reach an age of over forty unless he lived an active country life.[88] This was because city life had weakened the nerves creating a downward spiral of health.[89] Overall, what emerged was a broad juxtaposition between the natural environment of the countryside and the artificial one of urban areas. In such accounts, modernity itself constituted a threat to man's health as the demands of urban life directly challenged the intended functions of the 'natural' body.

Care of the stomach became particularly important when urban communities were threatened by epidemic diseases. A weakened stomach might render the body particularly susceptible to such outbreaks. A popular explanation for cholera transmission was that it entered the body orally and was dispensed outwards throughout the body from the digestive system. Accordingly, during the cholera outbreaks of the 1830s, some authorities warned that attacks were mostly made upon individuals whose modes of living were inclined to the induction of stomach and bowel derangement.[90] Care of the digestive organs came to be considered as a potential safeguard against cholera. A later account stated that the most important consideration to make when travelling to regions where cholera epidemics existed was to be fearless, as cholera stemmed not just from a derangement of the stomach but also from a weakness of character and will.[91]

Epidemic diseases, however, occurred only sporadically. Generally speaking, medical men were more interested in identifying day-to-day methods of soothing the stomach. Furthermore, gastric illness became perceived as a problem that was rapidly extending to all. During the eighteenth century, it had been commonly accepted that urban stomach problems resulted principally from the overworking of the mental faculties by middle- and upper-class intellectuals leading sedentary lives. Over-thinking was observed to have a weakening effect on the stomach, as the complex workings of internal sympathy resulted in nervous energy being over-expended in the brain at the expense of the digestive organs.[92]

The detrimental consequences of a sedentary lifestyle on the health of this particular social group certainly persisted as a relevant concept. A telling example of this can be found in reports of the suicide of the famed mathematician Professor James M'Cullagh (b. 1809) in Dublin in 1847. It was suggested that he had long enjoyed good health given that he was a man with literary habits, but in 1847 was suddenly and unexpectedly diagnosed as suffering from chronic dyspepsia. During the post-mortem enquiry made on his corpse, a physician explained that M'Cullagh had been working particularly hard. As a result, he had begun to suffer from paranoid delusions and had been advised to give up his mathematical studies. His apparent depression was blamed not on mental illness, which M'Cullagh had been reassured that he did not have, but instead on melancholy stemming from dyspepsia, which appeared to have originated from an over-application of the mind to an especially difficult mathematical problem. This, it was suggested, had encouraged the patient to neglect his bowels and to overindulge in strong green tea.[93]

Despite the persistence of concern regarding sedentary lifestyles, linkage between urban life and gastric illness became far more complex throughout the nineteenth century as Britain's towns and cities expanded. Gastric problems were identified as increasing as these urban areas sprang up and expanded throughout Britain. The overtaxing of the intellect in the sedentary life of the middle and upper strata of society became less of a primary concern when applied to the nineteenth-century urban context. In fact, medical authors began to associate stomach illness with a far wider range of occupations and lifestyles identifiable throughout the whole social scale. What emerged was a concern that a predisposition to gastric problems was spreading to all classes. As Watson Bradshaw claimed in 1864:

> From the gorgeously berobed potentate on his gilded throne, attended by his parasites, to the most squalid mendicant clad in rags and surrounded by destitution, this

> wonderful and curious malady discovers itself under an infinity of forms, pervading every ramification of society, remorselessly, unheedingly.[94]

There also emerged a distinct correlation between stomach problems and the physical labour of the nineteenth-century British towns and cities. In 1830, *The Times* noted that dyspepsia was highly prevalent in the factory districts and that, via nervous sympathy, indigestion had resulted in the loss of teeth amongst many young women residing in these areas.[95] A journalist for the *Manchester Times and Gazette*, writing two years later, was also concerned that many weavers who threw the shuttle by hand were being forced to retire from their jobs as a result of ill health brought about by the constant leaning over the stomach which this occupation entailed.[96] The abdominal health of handloom weavers, most notably those of Spitalfields in East London, continued to cause concern well into the 1850s. It was claimed in the *Era* in 1856 that 'constant working at the loom, the heavy weight falling continually on the chest, produces a large depression, almost a cavity at the pit of the stomach, which, of course, in its turn, produces dyspepsia and disease'. As a result of this, it seemed that no one in Spitalfields looked healthy, vigorous or happy.[97]

The business classes too became perceived to be liable to dyspeptic attacks. These professionals were also entangled in the constant whirl and rush of business, catching a midday meal only if they had time and being in the habit of drinking stimulants in order to set them up for work. These habits were especially notable at times when they were most harassed and worried about business. Yet it was during these periods of financial worry that they would be most likely to exhaust their nervous systems by eating and drinking too little whilst simultaneously working too hard. Worse still, even if wretchedly ill, they were likely to shun medical advice, fearing that they might be told to rest.[98] The physical nature of such office work was also postulated as a likely cause of dyspepsia. Those employed in offices were thought to risk damage to their stomachs due to the stooping position in which they wrote, a posture believed to interfere mechanically with the stomach's various actions.[99]

The wealthier classes, too, continued to be described as liable to dyspepsia, not least because these were the people most likely to reap the benefits of an advanced industrial economy, and who would have plentiful access to its associated luxuries.[100] Corsets, for instance, were a luxurious item thought to cause severe abdominal problems, and even death.[101] In a case reported in the *Hull Packet and Humber Mercury* in 1831, it was thought that the combination of a full meal and tight lacing had led to one girl's stomach becoming distended, which had disrupted the diaphragm and stopped air entering the lungs. The effect of the corset's impact upon the stomach combined with a crowded room, mental emotions and a dance had caused the tight-laced lady to faint, resulting in her premature death. Betsey Harris, a twenty-two-year-old from Middlesex, was detailed as having died in the same year due to the distension of the stomach and intestines with fluids and food. The cause of this was ascribed to tight lacing.[102] Similarly, in 1844, Caroline Koble's death in Chelsea was also blamed on the pressure of the stomach acting on the heart combined with pressure on the curvature of the spine which had thrown her heart out of its natural position. The stomach, in this case, was noted to have been particularly full due to the victim having eaten a hearty meat breakfast and supper that day.[103]

It is worth noting that a far wider range of problematic luxuries became identified. For instance, an article published in *The Times* in 1882, entitled 'Reminiscences of an Old Bohemian', claimed that:

> When he [the old bohemian] mentions that he has been a sufferer from dyspepsia, we really cannot wonder, and the less when we hear of the narcotics he indulged in. Smoking opium and swallowing hasheesh, inhaling laughing gas, sulphuric ether, and chloroform, mixing your Markobrunner with Erlanger beer, may introduce one to new sensations, but must be trying to the nerves and stomach.[104]

Overall, gastric problems such as dyspepsia became identified as having infiltrated British society to such an extent that they posed a threat to the health of all sections of society, from the working-class factory worker to those fortunate enough to be living in a state of luxury. The threat of chronic dyspepsia now appeared to be everywhere as gastric illness became constructed as an essential facet of the advanced state of civilization that Britain seemed to have arrived at. Influential medical theories which had prioritized the stomach within a highly complex system of nervous interactions added a distinct sense of urgency to this problem.

Managing British Gastric Illness

I have suggested so far that the stomach was awarded a prioritized location within the body and that gastric illness became perceived to be affecting increasing segments of British urban society on an unprecedented scale. Unsurprisingly, this generated a drive to target the stomach in order to treat many varieties of ill health. Yet the cause of poor gastric health in urban areas was considered to rest principally in shifting social and cultural factors. This resulted in the application of curative methods that focused on behavioural change in accordance with the new demands of civilization. As a result, the dissemination of medical knowledge of the stomach became directed towards enabling the individual to become capable of successfully navigating artificial urban obstacles. As Britain's transformation towards urbanity and civilization appeared unlikely to halt or reverse, a more plausible approach to the problem of communal gastric illness appeared to be to educate the public on how best to manage their new circumstances by paying close attention to their digestion.

A relatively limited role was accorded to cure by drugs.[105] Attempts to treat digestive complaints with drugs were invariably dismissed as hopeless for curing the problems of stomachs that had for too many years been exposed to detrimental predisposing influences. In the 1820s James Johnson claimed that more harm than good was being done by the endless varieties of medicines on offer for the dyspeptic. Instead, he suggested that nature itself would provide cure if assisted by the application of medico-scientific knowledge of the gastric system. He warned that 'flying to medicine at once' was a great error, although it was one committed daily by British practitioners and patients.[106] Robert James Culvervill, an early nineteenth-century medical practitioner, in his popular *Hints to the Nervous and Dyspeptic* (1837), insisted that great injury had been done to many stomachs, and therefore the bodily constitution, by the 'thousand and one panaceas for this disease [indigestion] that can be imagined'. Overall, he deemed medicine to be useless if the bad habits of living that acted as causative factors were persisted in.[107] This became related to a culture of therapeutic nihilism gaining increasing influence amongst medical professionals which facilitated the rejection of drugs as a tool for soothing the stomach by many physicians.[108] Mr Stomach, for instance, devoted a disproportionately large section of his memoirs to extensively detailing the evils that he had suffered under various prescriptions. In one edition, this was spread over thirty-seven pages.[109] In 1891, John Dewar, author of the *Red Cross Handbook on Dyspepsia* (1891), went so far as to advise his readers to 'throw physic to the dogs' and to 'use instead a little common sense in diet'. In his view, the public were half poisoning themselves with medicines.[110] It is also worth noting that travel was a further curative suggestion, as it was claimed to increase appetite and augment the digestive powers. However, not everyone had the time or finance to travel, and it became frequently described as the 'rich man's cure'.[111]

It was instead through dietary education that the digestive problems of urban environments could be more appropriately targeted and managed. At worst, the nineteenth-century urban environment was perceived to be a place of chaos and barbarity which broke long-standing traditions of cities acting as carriers of civilization and locuses of development of human culture.[112] The problems that emerged in British industrial cities defied eighteenth-century confidence in the potential of human society to improve and advance in the context of the town. Disillusionment emerged with earlier concepts that urban society would act as an agent of liberty, prosperity and civilization.[113] Commentary on digestive health was an important expression of this.[114] Although urban citizens might, on the surface, have appeared to be victims of a social and physical environment that was very much out of their control, in the eyes of many this did not absolve them entirely of responsibility for maintaining their health. Hence, they were still suitable subjects for remediation. Furthermore, it was believed that social order could be better maintained through providing education on appropriate habits and values.[115] The city acted as a place where people were free to act as they pleased yet was simultaneously a setting where they might easily lose their way if behavioural patterns were not persuasively and subtly adjusted.[116] In effect, the British social body became viewed as an entity that needed to be saved from wasting away.[117] Hence, the stomachs of its population required especial care if Britain was successfully to manage its ongoing transition towards civilization and urbanity. Digestive education was therefore intended not only to regulate the individual but also British society itself. The importance of dispensing dietary education became perceived as paramount.

Occasionally, the removal of non-dietary causes was suggested, although this became rarer. For instance, at the start of the century, the naval physician Thomas Trotter (1760–1832) attributed the prevalence of stomach disorders in Britain to the fashionable reading of romantic novels, which he described as 'love-sick trash', by women, and to the popularity of German drama, the latter described as a 'hotbed of diseased sensibility' and as a 'poison as has no antidote on the shelves of the apothecary'.[118] For him, the pervasiveness of communal abdominal problems might be countered by the removal of these seemingly detrimental mental influences. However, practical attempts to strengthen the stomach through adjusting mental activity proved difficult to employ as a curative tool. It was inevitably pointed out that the mind could not be healthy while the body remained diseased, and that a supply of books could not produce any benefit, unless it was accompanied by a supply of bread, beef and beer.[119]

Yet a gradual acceptance of medical theory regarding diet took hold throughout the century as the field of dietetics became placed on a firmer scientific basis.[120] Broadly speaking, the main concerns of digestive literature were the quantities of food to be digested by the stomach; appropriate timing and distributions of meals; and the moderation of levels of food variety. One commonly held opinion was that animals had been created in such a way that they could live healthily on a scanty, precarious supply of food. However, it seemed that civilized man had become too used to having food that he liked on demand and to receive gratification from its taste. According to Abernethy, modern man was eating and drinking an enormous amount more than was necessary for his natural wants, filling his stomach and bowels with putrefying food, the elements of which would then be dispensed throughout the body via the nervous system. The consequence of this, according to contemporary sympathetic theory, was nervous problems which might manifest throughout the entire system.[121]

It is important to note that it was commonly believed that digestion could be impaired with too much good food, as well as by the over-consumption of foodstuffs of less value to the system. The stomach, it was believed, could be accidentally trained to become used to overeating, eventually developing to a size grossly out of proportion to the rest of the body. As John Henry Clarke claimed,

> the stomach becomes to the gourmand what the athlete's limbs are to him – capable of an amount of exertion beyond the powers of other men. But over-development is not good in any part of the body, be it muscle or stomach; and the gourmand pays the penalty before long.[122]

Appropriate meal distribution formed a second important element of nineteenth-century advice on maintaining a healthy stomach. To many observers, urban life had disrupted man's natural schedule. One regularly made complaint was that the removal of men from the countryside to the town had encouraged a general postponement of dinner until 5 o'clock, if not later, in urban areas, a practice deemed to be ruining the health of thousands of city dwellers. One exception to this rule was Manchester, a city whose communal practice of dining at 1 o'clock was identified as being particularly beneficial for urban health, prompting claims that it should be introduced in every British town and city.[123] However, this raised alarm that it would impact negatively upon industrial productivity, provoking one anonymous Mancunian to argue in response that in fact:

> A Manchester man is never drowsy after dinner; he does not sink to the level of a boa constrictor, and indulge in a cozy, sulky snooze after eating; his motto is *simper vigilans* – wide awake; he knows nothing of dream-land; he cares nothing about fairy visions. He positively jumps up after eating a pound of beef-steak, and goes to his ledger as if nothing had happened. The Manchester stomach is *sui generis*; it is no more embarrassed by feeding than a steam boiler. *O dura mercatorum ilia!*[124]

The theme of natural meal distribution remained pertinent amongst both specialist and non-specialist literature on digestion. In the 1850s, Irish physician Arthur Leared (1822–79) observed a common practice of small breakfasts being consumed to allay the morning appetite but which ultimately proved insufficient for the body. He believed that the 'long unemployed organ' would secrete excessive mucus which underwent chemical changes at variance with digestion. The human stomach was naturally intended to have intervals where it would rest between episodes of digestion. However, for Leared, dinner, tea and supper were often being eaten unnaturally late and, to worsen matters, in quick succession. The nineteenth-century British stomach could not cope with this. Hence it became identified as one of the leading factors facilitating the breakdown of communal health.[125] Tellingly, an article entitled 'Tired Stomachs', published in the *Dietetic Reformer and Vegetarian Messenger* in 1881, warned that 'few persons of adult age have not already tired out their stomachs, or are in the process of tiring them out'. The stomach needed time to 'fold itself up, lie down and go to sleep' in-between meals as the organ was susceptible to fatigue.[126] Reassessing and regulating the habits of meal distribution in urban areas was a necessity if the British citizen was to live in accordance with nature's laws.

Thirdly, the increased levels of food variety available to the British stomach proved to be a concern. Modern life had encouraged global communication, one important consequence of this being an increase in the number of foods readily available for consumption. As discussed above, there had long existed an idea that natural food supplies had been placed around the world that were best suited to those who lived there. Yet new forms of transport including canals, railways and boats were bringing a highly refined and varied cuisine to all sections of British society, which the stomach was expected to digest. Dietary articles imported from entirely opposite climates were now being consumed in one single meal.[127] The availability of such a complex diet proved to be particularly troubling because it essentially formed a radical departure from the needs of the average British constitution. As a warning, it was stressed that most animals ate monotonously, urban man persistently defied the laws of nature by scarcely eating two similar meals in one day alone. Yet his natural digestive power was failing to keep pace with this change in eating habits.[128]

What, then, was the most appropriate form of natural diet for the British stomach? Some medical writers attempted to make substantial claims based upon their alleged knowledge of primitive man. The esteemed physician Thomas Lauder Brunton made a lot out of the subject of prehistoric strawberries as the seeds of these had been found in a body believed to date from prehistoric Britain. They must therefore be a good source of food.[129] Generally speaking, however, authors providing advice on digestive health eagerly expressed views on what substances were *not* good for the stomach rather than what were, due to them seemingly being deemed unnatural. Broadly speaking, it was condiments such as salt, vinegar, sauces and spices that were considered the most harmful to the British stomach, alongside stimulants including tea, coffee and alcohol. The lower animals were observed never to consume these as they contained no nutritive qualities. Their consumption complicated the natural work intended of the digestive organs, and represented a remarkable departure from the physiological needs of the animal world.[130]

Salt, for instance, was persistently dismissed as a mineral substance, being wholly innutritious and indigestible.[131] When taken into the stomach, it was seen to have negative consequences no matter how healthy the organ might have been before its consumption. Once absorbed, it would be dispensed throughout the body despite it holding no visibly positive internal contribution. Its long and habitual use seemed to result eventually in permanent organic sensibility of the stomach, which in itself caused general problems throughout the body. Salt addiction was even feared to be possible as the sudden stoppage of salt digestion would leave the organ in a severe state of depression.[132] Stimulants such as tea and coffee were also targeted. Their introduction had produced important changes in the customs of the European nations which to some constituted a profound revolution in dietetics.[133] Tellingly, the widespread habit of tea drinking was seen as especially problematic in regions of Ireland, and attracted considerable comment in the 1870s and 1880s when a tea epidemic apparently ravaged parts of the west of Ireland.[134]

Overall, however, it was alcohol that became most frequently attacked, not least because its impact upon the stomach's health was the subject of attention within the socially prominent temperance campaigns.[135] Some anti-alcohol advocates are reported to have attempted to impress the negative effects of intemperance on the audience attending their meetings by bringing along preserved, dissected stomachs. One prominent temperance advocate, Dr Sirder, to the apparent amusement of the audience, inflated one of these stomachs with his breath whilst speaking at a public meeting, and then asked the onlookers if they thought it was possible to put a pound of rump steaks and four pots of beer into such a small cupboard as that.[136]

Alcoholic indulgence was blamed for having a wide range of sympathetic consequences. As one journalist surmised:

> introduce spirits into the stomach, and only a portion is decomposed; the rest flies mostly to the liver and the brain, and then leaves the drunkard as it enters him. Tap the brain of a drunkard, and you will obtain pure spirits, a fact which accounts for spontaneous combustion.[137]

A further author, quoted in the *Manchester Times and Gazette* in 1842, went so far as to suggest that there existed three companions who one should be kept on good terms with: the wife, the conscience and the stomach, as intemperance produced domestic misery, wretchedness and premature death.[138]

Especially close links were established between alcohol usage and the nervous system. It was thought that regular alcohol usage led to the excessive expenditure of nervous power, as it overexcited and overtasked the stomach. Furthermore, its impact upon those living in urban, modern life seemed fundamentally different from previous eras. As one contemporary stated:

> We ulcerate the stomach to render it more capable of its functions! And what is the result of all this? Why that indigestion is become a national disease. The athletic husbandman, whose frame, in former years, was braced with nerves of iron, and who laughed at the weakling who talked of being nervous, now, from drinking ale and cider, trembles like an aspen leaf; and this sturdy rustic, who, in the days of our fathers, never felt that he had a stomach, now goes to the druggist for carbonate of soda, or keeps in his bed-room a box of antibilious bills!

On the idea that indigestion was a new disease that the labouring classes were particularly prone to, he claimed that

> on hearing a youth complaining of being nervous, an old woman, the other day, exclaimed, 'Nervous! nervous! People had no nerves when I was young!' They had what was better. They had nerves in a healthy state, and therefore they were never reminded by diseased tremors, that they had any nerves at all.

He went on to claim that savages who consumed no stimulants had virtually no diseases unless catastrophic events such as famine intervened. For him, 'our strong and wholesome ales and ciders, as they are called, our potent wines and cordials, as they are puffed, instead of bracing, have shaken, the nerves of the nation, and made us tremble at a shadow'.[139]

In 1847, temperance advocator Peter Burne depicted the impact of alcohol upon the human stomach using a series of illustrations which mapped the development of a range of stomach diseases which progressed through various stages of ulcer of the stomach, cancer and so on.[140] While the physiological accuracy of these images of the stomach is questionable, what they exemplify is the concept that the stomach was a useful resource employed to address a wide range of issues relating to nineteenth-century British society. In this instance, alcohol was weakening the national stomach. Burne depicted alcoholism as a specifically British problem, stating that the country's residents, whilst priding themselves on an unprecedented degree of knowledge and intelligence, sacrifice these efforts by their use of beer and ale as well as retarding the progress of society.[141] Of particular alarm to him were the rising levels of alcohol consumption throughout the country.[142] His work presents a mingling of anxiety over health, nationality and social progress. Hence, by utilizing physiological knowledge of the body, he intended to provide information designed to encourage the alteration of both individual and collective behaviour in the context of an underlying sense of foreboding for the nation's future health. It also acted as a warning, intended to prevent further abuses of the body. The diseases of the stomach are presented as occurring in a downhill sequence, with one disease metamorphosing into another. Overall, he is less interested in cure than prevention, with the added bonus of solving perceived problems of social morality.

The above example fully illustrates the primary points raised within this chapter. The stomach, as the primary organ of the body, was focused upon due to its pivotal function in terms of both individual and national health. Accordingly, the moral causes of illness became prioritized. Hence, cures for a disorganized stomach lay primarily within rectifying socio-moral behaviour. Recommendations for food consumption intended to maintain adequate levels of digestive health were therefore heavily intertwined within an overlapping debate between concepts of the importance of the stomach to bodily health; the fragile and complex relationship between the body, civilization, modernity and nationality; and the employment of physiological knowledge by physicians to complex social forces. Furthermore, this gave physicians a crucial role in maintaining public health in the context of the Industrial Revolution. By utilizing theories which prioritized the stomach and digestion, the medical profession was able to assert the validity of its common-sense interpretations of diet into an active, publicly visible form of social intervention.

2 THE ULCERATED STOMACH: GASTRIC DIAGNOSIS AND THE REORGANIZATION OF MEDICAL KNOWLEDGE, *C.* 1800–60

In 1857, William Brinton (1823–67) published the first monograph written in English devoted entirely to the medical phenomena of ulcer of the stomach. Brinton was an influential lecturer on forensic medicine and physiology at St Thomas's Hospital, London. Spread over two hundred pages, his *On the Pathology, Symptoms and Treatment of Ulcer of the Stomach* contains an unprecedented exploration of the disease's pathology, its incidence, detailed observations of its potentially fatal consequences, symptoms, aetiological theories and guidance on the most effective forms of therapeutics, concluding with an extensive set of case studies.[1] What is remarkable about the text was that such a detailed body of work could be published devoted entirely to a complaint that is virtually unheard of even thirty years earlier. In fact, prior to the 1820s, ulcer of the stomach had been discussed only occasionally, even within medical texts on the stomach.[2]

Between the 1820s and 1850s, a systematic conceptual framework was rapidly constructed around the complaint which facilitated the general acceptance of ulcer of the stomach as a unique, distinguishable disease entity. Brinton's work represents one successful outcome of pathological enquiries which aimed to understand the problems of the stomach through close scientific examination. Throughout the early nineteenth century, pathological anatomy elevated itself into an increasingly influential medical speciality which expressed a determined interest in forging new ways of knowing the stomach. This was made possible due to the essentially organ-focused approach of the discipline.[3] While clinicians did not abandon advice based upon prevention and environmental adjustment, texts such as Brinton's were illustrative of the rising influence of a more secular, scientific approach to gastric illness by the mid-nineteenth century. The diagnosis of ulcer of the stomach became stabilized and discernible from a complex array of gastric diagnoses available to early nineteenth-century practitioners, many of which were incorporated into the vague, overarching umbrella term dyspepsia.

Of course, the formation of clinically precise diagnoses was by no means a unique process to Britain. Anatomical exploration permeated medicine throughout Europe and America in this period. Nor was its gaze directed solely towards the stomach. All of the major body organs came under the scrutiny of the anatomist. For instance, the pulses of the heart were famously rendered observable by the introduction of the stethoscope in 1816.[4] Yet what is notable is that one of the most prominent British contributions to the discipline was the elucidation of knowledge regarding the stomach. The heightened cultural sensitivity directed towards the organ encouraged the formation of more precise definitions of gastric illness.

Knowledge of the Stomach

Throughout the early nineteenth century, a new ethos of medical professionalism brought about attacks upon a perceived vagueness of medical knowledge of the abdomen.[5] Claims were persistently made that the large quantities of books and articles available to both physicians and patients on the complex subject of the stomach were in fact increasing clinical uncertainty.[6] In part, this was a consequence of diagnosis and treatment being heavily dependent upon the individual views of the particular author, rather than being reliant upon a shared, systematic method of observation and description. Practitioners were therefore faced with a bewildering number of diagnostic and therapeutic options when presented with a dyspeptic patient due to there existing a wealth of differing, often contradictory, opinions and clinical recommendations.

A central part of the problem was that the stomach was hidden deep inside the body, which enabled opportunities to observe it in its living state to occur only rarely. The stomach remained largely inaccessible to external observational technologies.[7] In 1855, George Budd (1808–82), of King's College, London, complained that

> the stomach was not visible, could not be explored by hearing in the same manner as organs within the chest, its outline could seldom be felt like the liver or spleen, and its secretions were not easily poured out as in the case of the kidney.[8]

The stomach remained elusive and recalcitrant, evading the techno-diagnostic techniques that were meant to be acting as the bedrock of modern clinical examination. It also seemed liable to becoming affected by a wider variety of conditions than other organs, but these myriad conditions shared a remarkably similar repertoire of symptoms, making precise diagnosis profoundly problematic. To complicate matters further, a standardized physiological model of the stomach, as such, was not in existence. Following death the organ took on a variety of appearances, even when it had not been afflicted by disease. The preservation of the stomach of a corpse was dependent upon the general health of the cadaver at the moment of death, time of year, manner of death and the freshness of the corpse.[9] A fresh, healthy stomach suitable for pathological research could therefore rarely be guaranteed. Hence, the stomach remained difficult to observe and analyse in both its living and pathological states.

It is unsurprising, given this context, that diagnostic terminology remained vague, and was applied loosely in the clinical setting. The first known reference to dyspepsia had appeared in 1661 when it was described simply as 'imbecility of the stomach'.[10] It persisted as a distinct disease category despite the decline in usage of the humoral concepts from which it had emerged, although its symptoms became increasingly subsumed within models of the nervous system. By the nineteenth century, physicians commonly applied dyspepsia as a vague, umbrella term which, in its broadest interpretation, translated as indigestion, but which would undoubtedly have incorporated what today would be recognized as problems including gastric ulcer.[11] Even the stomach expert James Johnson remained sceptical about his own ability to provide truly accurate interpretations of the term 'indigestion', complaining in 1827 that

> this malady, or rather abstract of all maladies, is, in itself, such a Proteus – arises from so many different causes – assumes so many different shapes – produces so many strange and contrary effects, that it is almost as difficult to give it a name as to describe its ever-varying features.[12]

This predicament inevitably rendered decisions on appropriate therapeutic intervention problematic, which partly explains the persistent recourse to preventative health advice focusing on behavioural adjustment outlined in Chapter 1. Fears were constantly raised that texts on the stomach assigned contradictory forms of treatment for the same gastric symptoms, meaning that the young medical practitioner attempting to learn from them would soon discover, upon his entry into practice, that he would make numerous diagnostic and therapeutic errors. This held negative implications for perceptions of the work of the physician, not to mention for the well-being of the patient.[13] Tellingly, William Stokes (1804–78), esteemed physician at Meath Hospital, Dublin, and the leading authority in anatomical diagnosis in Ireland, complained to the *Medico-Chirurgical Review* in 1834 that:

> When you open a book on the practice of physic, and turn to the article dyspepsia, one of the first things which strikes you is the vast number of cures for indigestion. The more incurable a disease is, and the less we know of its treatment, the more numerous is the list of remedies, and the more empirical its treatment. Now the circumstances of having a great variety of 'cures' for a disease is a strong proof, either that there is no real remedy for it, or that its nature is very little understood.

He depicted this as a problem requiring urgent resolution. For him, until gastric illness became more precisely defined, danger existed that the patient might be suffering from a different illness than treatment was being provided for. This predicament was claimed by Stokes to be being confirmed daily by pathological anatomists who regularly dissected the corpses of patients who had been given inappropriate therapy with fatal consequences.[14]

Medical Reform and the Stomach

Complaints regarding the vagueness of knowledge of gastric illness were raised against a backdrop of dramatic theoretical and structural changes in western medical practice.[15] One of the primary aims of the late eighteenth-century Parisian School of medicine had been to assert the value of localized, anatomical ways of exploring the body. This formed an essential part of wider campaigns to replace long-standing traditions of diffuse, constitutional physiology. Furthermore, the aims of medical reform were focused upon bringing forth changes in the structure of medical education, establishing the hospital as a locus of medical activity and enhancing the status of the field amongst the general public by eliminating quack medicine and establishing in its place new principles of scientific enquiry. To achieve this, medical exploration became increasingly focused on organs. Analysis required the anatomical examination of dissected cadavers in order to determine the relationship of pathological anomalies with the symptoms that the patient had suffered from prior to death. The stomach came to occupy an important location in the ideologies and methodologies surrounding the move towards pathological anatomy.

In France, the relationship between localized abdominal problems and the general nervous constitution was expressed most vigorously by the prominent French physician François Joseph Victor Broussais (1772–1838), who was especially interested in localism. He proposed that overstimulation of the gastro-intestinal tract by external factors including cold air, food and drugs produced a local irritation in the stomach which was likely to transform into general, bodily inflammation. Broussais's trust in his principles of gastric irritation was such that he denied altogether the existence of disease specificity. Even the causes of smallpox, syphilis, measles, hypochondria and cancers became reducible in his model of the human body to gastric inflammation. Broussais famously warned that it was 'the destiny of the stomach always to be irritated'.[16] In France, for a while, gastroenteritis became seen as the key disease to be treated for all medical complaints, although ideas regarding its significance were to be superseded by concern over the widespread prevalence of liver complaints within the country. Yet this pronounced interest in the stomach proved to be easily transferable to

British medicine, given the enhanced levels of medical scrutiny that the organ was subject to in a country where gastric illness held strong cultural resonance.

The influence of the ideals of Parisian medicine rapidly spread throughout Britain following the end of the Napoleonic wars. By the 1820s, the country's medical world became aware of an exodus of many of its brightest students to France.[17] Upon their return, an influx of a new scientific approach was witnessed which brought forth new approaches to understanding the stomach and which strove for higher levels of scientific integrity.[18] In Britain, surgeons and pathologists remained influenced by Hunterian principles, and expressed less enthusiasm for inflammation than their French counterparts. This was because Hunter's work was held up as the ideal of scientific medicine in this period, and Hunterian principles of sympathy that stressed complex systems of interactions between the stomach and other body parts offered a firm conceptual grounding for surgical work.[19] They became applied practically by pathologists and surgeons to express their burgeoning scientific prowess and the authority of their discipline within the medical profession as a whole.[20]

John Abernethy, for instance, who, as we have seen, placed especial emphasis upon the stomach, envisioned a much stronger role for the surgeon than his traditional occupation as provider of therapy for external bodily complaints. By stipulating that localized internal gastric disorder caused general bodily disturbance, Abernethy attempted to carve out a new role for the modern surgeon which involved making enquiries into the nature and treatment of medical complaints inside the body, and thereby aimed to establish a new professional niche.[21] Emphasizing their ability to understand the stomach therefore allowed surgeons and pathologists to direct their gaze towards the inner body, rather than restricting themselves to the external alone.[22] This also helped to bridge the gap between medicine and surgery. If successfully achieved, surgeons would be equipped with far greater authority in managing the perceived epidemic of chronic gastric illness in Britain which would simultaneously heighten the standing of their discipline. Yet public attitudes to pathological anatomy were tainted by accusations of bodysnatching and, from 1832 onwards, the utilization of corpses of paupers.[23] Hence, the anatomical intrusion into the stomach served important professional purposes as well as diagnostic ones. One consequence of success in this area would be to enhance the status of their discipline amongst the general public.

It is therefore unsurprising that an increased surgical interest in the stomach is observable in much early nineteenth-century medical literature. For instance, John Howship (1781–1841), author of *Practical Remarks on Digestion* (1825), warned that practitioners and patients were frequently misdiagnosing medical conditions due to the misleading symptoms produced by the complex sympathetic interactions of the nervous system. Notably, Howship was a surgical lecturer at Charing Cross Hospital, London, who maintained strong connections with French medicine as an honorary member of the Société pour les Sciences Naturelles et Medicales. Like many of his contemporaries, Howship was also highly concerned with bridging the perceived gap between medicine and surgery.[24] Hence, his agenda whilst compiling research into the stomach was to contribute to a wider professional project which aimed to transform the practice of British medicine and to unite more closely the disparate fields of medicine and surgery. Further specialist publications followed, one of the most notable being James Johnson's *Essay on Morbid Sensibility of the Stomach and Bowels* (1827), a text discussed more fully in the previous chapter. Johnson, who was also trained in Paris, was particularly concerned with reforming medical practice and shifting power into the hands of a new type of educated medical professional.[25]

Given the emphasis placed upon the stomach in nineteenth-century British medical and popular thinking, it is unsurprising that advancing scientifically defined knowledge of the organ became an important facet of the processes of surgical and pathological professionalization. The organ acted as an especially suitable target of observation within an era witnessing a reorganization of medical knowledge and practice. Furthermore, the form of knowledge produced relating to the stomach was intended to differ vastly from that given within the more popular texts on digestion in that it depended less upon the agency of the patient to negotiate his or her own health and instead required the specialist insight of the modern medical professional.[26]

Ulcer of the Stomach

It is in the context of reform in British medicine and alarm over the vagueness of pre-existing knowledge of such a central organ that new diagnostic categories of gastric illness were developed. This process differed from the clarification of diseases in other inner organs. Discussion of the heart tended to focus upon aural inspection, utilizing examination of murmurs and tremors.[27] The liver was depicted anatomically, but was never imbued with the same sense of importance within the nervous system as the stomach.[28] Similarly, few British texts emerged entirely devoted to the kidneys or bowels, not least because these areas did not seem to have the same pathological complexity, or cultural resonance. The uterus was also imbued with high levels of sympathetic connection. Yet it was more likely to be discussed in terms of its relation to non-medical phenomenon such as childbirth, or analysed to explain constitutional differences between the male and female sexes, rather than to help resolve a perceived national health problem.[29] It was the stomach which became persistently conceptualized as one of the key organs of pathological enquiry in Britain.

John Abercrombie (1780–1844) provided the first full pathological description of ulcer of the stomach in 1828. Abercrombie was a renowned Edinburgh physician and is remembered principally as being the first medical person to treat neuropathology as a distinct medical specialism.[30] What was strikingly different about his approach from that of earlier generations of physicians was his allegiance to ideas that if medicine was to become truly scientific, then less attention needed to be paid to theories and systems which he dismissed as 'the assumption of principles which are altogether gratuitous and imaginary'. For him, such speculation had been produced with no use of actual physiological observation. Accordingly, he dismissed them as holding little or no factual foundation despite the great confidence placed in them as established principles. Abercrombie suggested instead that more precise bodily enquiry undertaken during post-mortem examinations would ultimately provide a more exact record of pathological facts.[31]

Unsurprisingly, Abercrombie subjected his observations on the stomach to a far higher level of scrutiny than both his contemporaries and predecessors. He began his *Pathological and Practical Researches on Diseases of the Stomach* (1828) by completely dismissing the existing body of knowledge on the gastric tract, claiming that

> there are few points in medical science which have undergone more discussion than affections of the stomach; and yet, it must be confessed, that when we come to investigate the subject, according to the rules of pathological induction, we find little that is satisfactory.[32]

He went on to suggest that one of the reasons for this was that, despite the great discomfort generated by chronic gastric illnesses, they tended not to be fatal. This therefore allowed for fewer opportunities for pathological investigation unless a patient had happened to die of another disease and if the stomach had happened to have been explored. The organ therefore presented Abercrombie with a wide field for speculation, conjecture and empiricism and the potential establishment of firm principles relating to the diagnosis and treatment of its complaints.

The influence of an organ-based approach is clearly discernible within Abercrombie's work. One of the novelties of his research lay in its demarcation of the different parts of the organ into highly specific regions. The stomach itself effectively became dissected and reduced to a more clearly understood organ with clearly defined areas. In his writing on the organ's diseases, he divided the stomach into the peritoneum, the muscular coat and mucous membrane. He then separated discussion of medical complaints into inflammatory affections, organic affections and functional affections. This categorization facilitated close discussion of problems such as acute gastritis, a disease observed to have been described by almost all medical writers, but which had not yet been fully depicted using the tools of pathological examination. Crucially, Abercrombie located this as being a problem of the mucous membrane alone. He then discussed a range of organic diseases resulting from complications such as the thickening of the coat of the stomach, and chronic peritonitis. What was unique about Abercrombie's approach here was the close attention paid to the problems of precise regions of the stomach. Notably, he observed that the same disease entity might produce an entirely different spectrum of symptoms depending on which part of the stomach wall it was located in.

What Abercrombie provided was a close pathological account of the stomach, which recognized that while functional problems such as dyspepsia did exist, the medical problems of the stomach required closer attention as many of them were connected with clearly definable areas of the organ. Unsurprisingly, he dismissed the widespread utilization of the vague phrase dyspepsia as being at variance with the principles of sound investigation. The stomach became presented as an area of far greater complexity than had previously been recognized, and became depicted as deserving of sophisticated diagnostic categorization if its ailments were to be successfully mastered. The accumulation of this knowledge also served to show off the prowess of new pathological techniques.

Ulcer of the stomach emerged from Abercrombie's refined diagnostic system as one of the most significant categories of stomach diseases which pathological anatomy had been able to elucidate. His description of the condition stated that:

> We have every reason to believe that the mucous membrane of the stomach is liable to inflammation in a chronic form, which often advances so slowly and insidiously, that the dangerous nature of it may be overlooked, until it has passed into ulceration, or has even assumed the characters of organic and hopeless disease.[33]

Abercrombie portrayed ulcer of the stomach as a disease with a very slow progress, which typically developed for a considerable length of time although its syndromes might periodically re-occur until it eventually became a permanent and extensive disease. Yet this was a complaint whose symptoms often remained unnoticeable until it had developed into a more serious problem such as perforation or haemorrhage. He also identified three broad categories of ulcer: a small ulcer with some loss of substance and where there was often relatively little damage to other areas of the stomach; ulcers similar to these but which were complicated with thickening and induration of the parietes of the stomach; and extensive irregular ulceration of the inner surface of the stomach, generally complicated with thickening and induration of the coats and fungoid elevations. He went on to describe the different pathological states of ulcers within the various regions of the organ. Following his establishment of these pathological observations as constituting a serious disease entity, he suggested that successful cure lay in halting the process of inflammation which he had observed and hoped would prevent ulceration from occurring before it had time to cause permanent organizational damage.[34] Yet the success of this depended on knowing what type of ulcer existed and where in the stomach it was situated.

Abercrombie's work became highly praised in professional journals. The *Edinburgh Medical and Surgical Journal* suggested that:

> To the medical reader Dr Abercrombie has rendered an important service, by putting him in possession not only of the results of his own observation, but of the illustrative and confirmatory facts derived from the collateral inquiries of others. To pathological science he has also performed a meritorious duty, in cultivating a field hitherto waste, in extending the bounds of knowledge and in correcting several erroneous notions.[35]

The *Medico-Chirurgical Review*, meanwhile, commended him for his:

> Philosophic cast of mind, and his ample opportunities for observation; his jealously of principles too implicitly received, and his distrust of authority too imperious to be useful; his patience in investigating every circumstance, and his prudence in explaining every fact; his dread of theory, and his fondness for induction.[36]

Overall, Abercrombie's new diagnosis resulted from shifting understandings of medicine and was coloured by a wider concern within Britain to reorganize medical knowledge of the stomach. Ulcer of the stomach emerged in the medical literature at this time principally due to shifts in the professional structure of medicine which aimed for higher ideals of scientific excellence.

Explaining Perforation

Historically, the medical profession have constantly constructed new diagnoses. It is therefore not enough to state that ulcer of the stomach was first properly recognized in the 1820s and then to assume that the medical profession began to utilize it diagnostically. Instead, particular conditions must have been in existence which encouraged the dissemination of it as an important disease entity. One particular variety of ulcer of the stomach – the perforating ulcer – became especially prominent in both medical and popular discourse as it became heavily entwined with debates relating to the role of gastric acid, poison and alcohol in causing structural change.[37]

By the 1830s, medical education began to act as a more effective vehicle for the transmission of newly accumulated professional knowledge. The decade saw a sharp increase in the number of pressure groups demanding that teachers and practitioners outside of the traditional power structure had more say in the management of medicine. New teachers, university chairs and academic courses testified to a rapid growth of interest in comparative anatomy.[38] This would have facilitated the dissemination of knowledge of the condition throughout the medical community. Yet other forces were at work which facilitated the general acceptance of ulcer of the stomach as a distinct disease entity. During the 1830s and 1840s, pathologists became especially intrigued by the perforated ulcer. The classification of this anomaly had often been problematic, as whether a hole discovered in the wall of the stomach was a result of post-mortem decay or ante-mortem disease had proven difficult to establish. The new ethos of internal pathological observation encouraged speculation into the potential of gastric acids to make considerable-sized holes, or perforations, in the organ's wall.[39] This debate had been stimulated by the relatively recent agreement that digestion was a predominantly chemical process, rather than one characterized by putrefaction or mechanical grinding as had been speculated in earlier periods. Prior to this, knowledge of what happened to food once it had been ingested and passed through the interior areas of the body had remained mysterious due to chemical digestion occurring in areas not easily accessed or visualized.[40]

The role of saliva in breaking down starches and the ability of the gastric juices of the stomach to dissolve food had been established during the eighteenth century by the French naturalist René Réaumur (1683–1757) and the Italian physiologist and natural scientist Lazzaro Spallanzani (1729–99) respectively.[41] Similarly, Edward Stevens (1755–1834) of Edinburgh had experimented on a person who had been in the habit of swallowing stones and hard objects for over twenty years, causing his stomach to become heavily distended. The sound of the stones colliding in his stomach could even be heard by passers-by, so it was claimed. Stevens convinced his patient to swallow perforated silver balls containing raw animal food, vegetables, leeches and other substances in various stages of mastication as part of an attempt to prove the dissolving effects of gastric fluid.[42] These new ideas were eagerly incorporated into nineteenth-century medical practice.[43] The prominent physician and biological chemist William Prout (1785–1850), for instance, argued that physiologists typically paid too much attention to mechanical or metaphysical explanations of human biology, suggesting instead that their work required the unlocking of chemistry's diagnostic and therapeutic potential. Notably, in 1824 he showed that the acid of the gastric juice was hydrochloric acid, and also made important distinctions between carbohydrates, fats, proteins and water, translating his findings for practical and clinical usage.[44]

As chemistry increasingly asserted itself as applicable to medical inquiry, problematic questions emerged relating to the potential of gastric acid to produce decay in the cadaver, and disease in the living human. This sharply inflected discussion of perforating ulcer of the stomach. Hunterian ideas of vitality had stipulated that the digestive juices could only alter the stomach's structure following the loss of its 'living principle' upon death, at which point the stomach would begin to digest itself. Hunter became especially concerned with the role of violent death in producing gastric decay, not least because many of his cadavers were hanged criminals. To prove his point, he conducted a series of investigations involving killing fish violently and examining their stomachs afterwards. He concluded that were that there were likely to be few dead human bodies in which the stomach had not begun to digest itself. Hunter's influence in Britain ensured that notions of natural decay were emphasized when describing pathological findings.[45]

Yet despite Hunter's influence, the possibility that the phenomenon of perforations within the stomach might in fact result from disease gained increasing validity.[46] The famed London physician Robert Carswell (1793–1857) stated in 1830 that

> it is not easy to conceive how any one could so far disregard or undervalue the influence of the gastric juices in occasioning certain organic injuries both before and after death; since it is the peculiar property of this fluid to change, and in a manner to reduce to their most simple elements, all animal substances subjected to its action.

Accordingly, Carswell suggested that ulcer of the stomach was certainly not a new disease, despite its recent appearance in the pathological literature, but had in fact been regularly described in earlier accounts masquerading under the description of 'softening and destruction of the mucous membrane'.[47] Yet the debate over the potential impact of gastric acid on the living stomach wall helped to consolidate the isolation of the perforating ulcer as an important and deadly disease entity. Furthermore, discussion of it within contemporary medical literature inadvertently helped to disseminate knowledge of ulcer of the stomach amongst contemporary medical men.

The perforating ulcer also became widely acknowledged amongst the general public when it became entwined with discussion of poisoning. The mid-nineteenth-century medical profession witnessed a fascination and alarm over murders committed by slow, gradual poisoning. Interest in the perforating ulcer heightened as the medical profession became more actively engaged with legal cases during the early nineteenth century.[48] Poisoning cases provided an ample opportunity for the medical profession to extend its social influence into the courtroom. Yet fears became raised by medical professionals providing evidence that the symptoms of poisoning might easily be confused with those of the perforating ulcer, leading to inaccurate testimony.[49] Discussion which ensued throughout the 1840s further helped to raise general awareness of perforating ulcer of the stomach, as evidenced by the publication of highly detailed accounts in medical journals, pamphlets and textbooks.

Tellingly, the influential medical jurist and toxicologist Alfred Swaine Taylor (1806–80) published a lengthy article on the perforating ulcer in *Guy's Hospital Reports* in 1839 and devoted considerable attention to it in his *A Manual of Medical Jurisprudence* (1844). Within these contributions, he expressed concern that perforations from ulcers and poison were being commonly confounded. In both scenarios, seemingly healthy people would rapidly develop similar symptoms including vomiting and violent abdominal pain soon after a meal. A sudden death would then typically occur. Taylor thought it necessary to ensure that practitioners were aware of the subtle differences between the two forms of death as there was a risk of miscarriage of justice occurring should a false accusation of murder be raised.[50] Accordingly, to distinguish arsenic poisoning from natural perforation, he stressed that the former's symptoms typically commenced around thirty minutes after a meal, rather than the latter's three to four hours, and that overall there was a more gradual pain in poisoning than in cases of natural disease.[51]

Similarly, Thomas Williamson, physician to the Leith Dispensary in Edinburgh, published a lengthy article in the *Dublin Journal of Medical Science* in 1841 highlighting the importance from the medico-legal point of view of distinguishing between the two types of perforation. Within this, he stressed that there seemed to be little doubt that in earlier periods the moral character of many innocent people had been sacrificed due to a lack of medical knowledge. He considered this to be a product of a limited awareness of the problem of perforative stomach disease. Williamson attempted to alleviate this situation by closely analysing the differences between perforations from gastric juice, ulceration and poison.[52] Set against this backdrop, it is unsurprising to find that the first pamphlet devoted to ulcer of the stomach, written by a physician named Edward Young and entitled *On Perforating Ulcer of the Stomach* (1849), was devoted entirely to this medico-legal dilemma. Within it, Young stated that the disease was in fact far more common than was often thought and that the urgent dissemination of medical knowledge relating to the perforating ulcer was necessary in order to provide satisfactory explanation for what initially seemed to be a suspicious death, alarm over which might have initially been raised by the physician.[53]

What the debate regarding gastric disease and poisoning inadvertently did was to encourage a more general awareness of ulcer of the stomach amongst both the medical profession and the British public. Certainly, the perforating ulcer became increasingly referenced in a wide range of contexts. Notably, contemporary temperance literature made increasing usage of this new diagnostic category as it seemed to add scientific weight to their cause. This increasingly socially prominent movement became eager to utilize medical ideas relating to the potential of drink to produce ulcerative disease by irritating the nerves of the stomach, not least because this seemed to provide scientific evidence which justified the aims of their campaign.

Post-mortem reports of ulcer deaths thought to have resulted from overindulgence in alcohol began to appear frequently in temperance publications. In 1841, the *Temperance Lancet and Journal of Useful Intelligence* published a letter from a physician who described a man who drank too regularly, and who, when wrestling with friends, had suddenly complained of abdominal pain and died the day after. The physician found the coat of his stomach covered in small pyloric ulcers. The sudden death caused by the perforation of one of these ulcers was attributed by the author to drink. Overall, it was utilized to draw attention to the latent, hidden problems that drinking might cause in a description that read:

> Oh that we could but reach every mind upon this truly awful subject. Here is a young man 'frolicking', not drunk ... No – he was 'comfortable', and within four days a lifeless corpse. And yet the organic disease above related must have been progressively accruing for years.[54]

In 1844, the *National Temperance Magazine* referred to the physiological experiments of William Beaumont which had observed that the inflamed and ulcerated appearance of the stomach was a likely result of excitement by stimulating liquors, which had overloaded the stomach with food, fear, anger and other substances that depressed or disturbed the nervous system.[55] In 1849, meanwhile, the *People's Abstinence Standard and True Social Reformer* also described ulceration of the stomach in an article entitled 'The Death-Bed of the Inebriate'.[56] Finally Reverend Benjamin Parsons (1797–1855), in his *Anti-Bacchus* (1850), claimed that 'the stomach, which is one of the most important laboratories of our frame, is injured as soon as this vile spirit enters it. It becomes, as actual observation has now demonstrated, inflamed and ulcerated.'[57]

Notably, Ralph Barnes Grindod (1811–83), the so-called 'medical apostle of temperance', regularly gave lectures to popular audiences on the advantages of temperance throughout the 1840s. He would take physical examples of dissected, ulcerated stomachs along to these meetings and present them as firm evidence of the negative effects of alcohol on the stomach.[58] In 1849, for instance, he exhibited a human stomach with a hole in it, claiming that excessive brandy drinking had caused an ulcer to perforate which had resulted in a sudden death. In this instance, it seemed that alcohol dispensed to a sailor as a medicinal cure had proven fatal. Initially, the surgeon who examined the patient suspected that corrosive poison had been consumed but was unable to prove his theory, instead concluding that the condition had been caused by ardent spirits. A debate raged in a local journal in Bath regarding this death for a number of weeks.[59] Such accounts clearly demonstrate the intrusion of new disease entities accrued from pathological enquiry into the public sphere, although these were still interpreted in the context of nervousness and education on preventative measures.

Overall, the role of chemical agents in producing structural damage to the inner surfaces of the stomach encouraged attention to be increasingly directed towards the perforating ulcer of the stomach during the 1830s and 1840s. These processes aided the acceptance of the isolation of the disease category as distinct from an apparent mass of vague diagnoses, and further helped to stabilize ulcer of the stomach as a recognizable disease entity. It also exemplified the ability of contemporary pathological science to distinguish new diagnostic categories and to present them convincingly to the general public which in itself served an important role in the professional reorientation of medical activity.

Therapeutic Optimism

Shifting processes of professional activity clearly encouraged the formulation of ulcer of the stomach as a coherent diagnostic category. Set against a backdrop of conceptual debates regarding the impact of chemicals within the gastrointestinal tract, perforating ulcers quickly became recognized as an important medical problem which was diagnostically distinguishable from dyspepsia. Knowledge of it rapidly spread throughout the public sphere. However, the accumulation of more precise diagnostic techniques did not imply an automatic improvement in therapeutic regimes. Curing gastric conditions even when more accurate diagnoses were available remained problematic. The *Medico-Chirurgical Review* had taken a particularly grim approach to this predicament in 1831, pointing out that the many diagnostic mistakes made by physicians when dealing with the organ did not actually result in much loss of life, as the diseases of the stomach would still remain incurable even if accurate diagnosis was achievable.[60] An excessive reliance on leeches brought about by the extreme, but popular, therapeutic methods of Broussais had seemed to enhance this mood of pessimism, as it raised alarm that such methods of gastric therapy were in fact potentially worsening the patient's illness.[61] Critics went so far as to claim that the most successful therapeutics for treating gastric conditions were in fact those which Broussais's followers condemned.[62]

What emerged in the case of ulcer of the stomach was a recourse to rest. Drugs were often rejected, or at least used sparingly, as they seemed to have an unpredictable effect on the stomach, particularly if the exact nature of the gastric complaint remained unclear, as it would have done in many cases.[63] Furthermore, it was feared that an excessive reliance on pills might result in a persistent culture of self-medication. A reviewer writing for the *Medico-Chirurgical Review* in 1828 was concerned that some members of the medical community were deliberately providing bad advice to patients in order to induce stomach disease, which would allow them to drum up more business. He used the analogy of an undertaker setting up a pharmacy next door to the building where he made his coffins, his main concern here being that 'people would be rather shy in going to the former, lest they should be carried to the latter'.[64]

Generally speaking, ulcer of the stomach remained perceived as untreatable. Yet occasional observations of perforated ulcers that appeared to have healed up and closed were noted, which was suggestive that natural healing might occur.[65] This remained speculative. Yet a rapid change in outlook occurred at the end of the 1840s as perforating ulcers became identified as in fact potentially curable, if subject to the right conditions. One of the earliest accounts of recovery was published in a highly detailed report in *Guy's Hospital Reports* in 1846. H. M. Hughes, a Dulwich physician, reported that he had been treating a twenty-eight-year-old domestic servant named Sarah who had for months been suffering with bilious attacks and severe stomach pains. Following one particularly strong attack of this, she was treated with opium and went on to spend fifty-two days in hospital. Hughes was convinced that Sarah had recovered from a medical complaint which matched all the symptoms of perforating ulcer of the stomach. Yet he was certain that his controversial report would come under criticism, not least because within it he was claiming that the patient had recovered from a disease that was considered incurable. He seemed to be describing the impossible. What he proposed was that he had witnessed the natural healing of a perforating ulcer encouraged by opium, rest and starvation.[66]

Sarah was readmitted four months later and this time died of a perforated ulcer. According to Edward Ray, the physician who treated her on the second occasion, she had failed to follow Hughes's strict dietary regulations, and her death was blamed on the consumption of fruit a few days prior to death which had reopened her wound. Despite the tragedy of the death, it provided an excellent opportunity to determine whether or not a stomach ulcer had in fact cicatrized during her first admission earlier that year. To much surprise, Ray's post-mortem examination seemed to confirm Hughes's controversial assertion of the possibility of a cicatrized ulcer, which was discovered to be around the size of a four-penny piece and which had extended through the fibrous structures of the mucous membrane.[67] Although recovery continued to be seen as rare, similar examples began gradually to appear within medical journals. For instance, a twenty-one-year-old Irish tailor travelling to America began to suffer from incessant vomiting attacks on his seventy-day voyage. Upon landing in Nova Scotia, he was reported to have spent a further six months vomiting. He procured a bed at the Massachusetts General Hospital and was prescribed an iron treatment to reduce the attacks. After almost a year of this, he died and was then discovered to have a stomach ulcer that had previously cicatrized.[68]

Cases such as these inspired increasing levels of therapeutic confidence. Notably, by the mid-1850s George Budd was claiming to be accurately diagnosing and treating ulcer patients on a daily basis. Budd had pursued his medical studies in Paris and attracted attention by writing on his research into the stethoscope as an acoustic instrument and his extensive research on cholera and scurvy. In 1840, he was elected professor of medicine at King's College, London.[69] It was here where he first began treating ulcers by applying physiological findings, suggesting that healing would be retarded by the constant churning motion that takes place during digestion. He maintained that the healing of the ulcer would in fact be stalled by the stomach's gastric juices as, although not injurious to the living mucous membrane, it could dissolve and remove the lymph surrounding the anomaly which was necessary to healing. For Budd, this explained how a small ulcer, which caused no obvious constitutional disturbance, might become such a serious, fatal disease. The therapeutic methods suggested by him therefore concentrated on halting substances from coming into contact with the organ in order to reduce the chances of irritating the ulcer.[70]

Budd also stressed that death would commonly occur following a meal when the solvent action of the gastric juice, or the distension of the stomach during digestion, were strongest. When the ulcer had eaten down to the peritoneal coat, the solvent action of gastric juice might exert itself and cause perforation. It therefore seemed necessary to avoid the writhing movements of the stomach's coat during digestion, lower levels of secretion of gastric juices and control the substances introduced to the stomach to those that were farinaceous, but nutritious, such as bread, macaroni, semolina, biscuit powder, flour and oatmeal. However, he still considered it to be the patient's responsibility to stick to these diets.

Budd's work represents a mingling of scientific methods of cure with a traditional recourse to prevention and the correction of detrimental socio-moral behaviour due to its impact on the nervous system. Budd explicitly referred to the varieties of preventative advice available, most notably that concerned with alcohol. He directly referred in his writing to the aforementioned diagrams of the ulcerated stomach of the drunkard from *The Teetotaller's Companion*.[71] He believed that the post-ulcer diet would only prove successful if the patient was of temperate habit, and if their nervous system did not require alcohol. Certainly, he was prone to showing publicly examples of stomachs of patients who had recovered but who had persisted in leading an intemperate life which had ultimately led to their death. Ardent spirits, he insisted, if digested over a number of days, led the mucous membrane to lose its smooth and healthy appearance, and its secretions to be diminished.[72]

What Budd and Brinton's work represents is the culmination of decades of debate regarding the pathological state of the stomach and the influence of digestive chemicals in producing disease. William Brinton's work was published soon after Budd's. Brinton claimed to have seen as many as 15,000 cases in St Thomas's Hospital within a period of only five years. This led him to conclude that the prevalence of ulcer of the stomach was generally underestimated by physicians and had been commonly misdiagnosed. For him, it was not a new disease, but was instead one that had been mistakenly considered to be rare, a problem to be overcome by the dissemination of knowledge of it as a curable, identifiable and serious illness. He went so far as to claim that the open ulcer was observable in 2.5 per cent of the bodies of people dying from disease, with evidence of cicatrization being found in as many as 5 per cent. Furthermore, he suggested that although attention had for decades been placed principally upon the problem of perforating ulcer of the stomach, only around 13 per cent of cases observed had reached this complicated stage. Ulcer of the stomach was therefore presented by him as a highly prevalent condition, but one which was latent throughout British society.[73]

Like Budd, Brinton also presented a particularly optimistic account of therapeutic possibilities, aiming to remove all local obstacles to the cicatrization of the ulcer and to assist the constitution in this process. However, he differed in his confidence in the use of medicines, in particular bismuth. Although he claimed to be ignorant of the exact processes by which this worked, he was careful to justify their use with their particular effects on specific symptoms. One of the main changes which he instigated was the introduction of the milk diet to provide cure for ulcer of the stomach.[74] Milk was considered to be especially nutritious, whilst being unlikely to harm the stomach in any way when used medicinally. The milk diet persisted as a common form of treatment well into the twentieth century.

Brinton's work provides a mingling of the old and new. Diet was still utilized therapeutically, but dietary recommendations were suggested based upon more scientific premises than those in the popular literature available. Closer analysis of the impact of diet upon the body's internal physiology meant that food articles were recommended for more precise scientific reasons. Furthermore, Brinton, like Budd, saw ulcer of the stomach as a condition more likely to be noticed, and to be better managed, within the hospital setting. It therefore relied less upon the autonomy of the patient to manage his or her diet at home, and required higher levels of professional medical attention than the management of dyspepsia. It was essentially a transitional approach, combining traditional gastric therapeutics with new research findings. Brinton's work also exemplified the ability of pathological enquiry to overturn approaches to illness of the stomach. Ulcer of the stomach therefore emerged as an especially common entity which had been previously unnoticed, and one which the scientific influence directing modern medicine had the potential to manage and cure successfully.

Conclusions

The early nineteenth century clearly witnessed a reorganization of medical knowledge that impacted upon the manner by which the stomach was conceptualized and managed. Pathological science and the rising ambitions of anatomists rapidly led to the construction of a body of knowledge which added complexity to the highly popular literature available on digestion which focused principally upon regimes of preventative health. This was made possible by the introduction of new ways of observing the stomach which involved closer analysis of the different sections of the organ and the different influences of diseases on those particular regions. Given the profound cultural anxiety surrounding the stomach in Britain, pathological anatomy had much to gain there in terms of heightened professional prestige and social acceptance by elucidating the problems of the organ. The accruement of knowledge of ulcer of the stomach, and associated claims to be able to manage and cure it, was therefore illustrative of the wider aims of the discipline of pathological anatomy.

However, it is important not to depict an overly optimistic picture of the achievements of the medical profession in this period. Gastric illness persisted as a problematic area of diagnosis and treatment, despite the vast amount of literature on the subject that appeared throughout the 1850s. In fact, it was suggested in 1862 that there now seemed to be in existence incessant varieties of stomach disease, each of which now needed to be fully understood.[75] Diagnosis remained difficult, and it was expected that the findings of chemical science would instead provide answers through its examination of digestion.

3 THE LABORATORY STOMACH: GASTRIC ANALYSIS IN AN ERA OF VIVISECTION AND FORCE-FEEDING CONTROVERSIES, *C.* 1870–1920[1]

The potential of medical science to assess accurately the medical complaints of the stomach was asserted in an unprecedented manner from the mid-nineteenth century onwards. Physiologists began to stress new ways of understanding the human body which claimed to be more precise than the work of pathological anatomists. This resulted from the fascination with the digestive system held amongst late nineteenth-century laboratory researchers. Of course, medical and scientific investigators had long aimed to understand fully the nature and functions of the gastric system. Yet, the retrieval of the contents of the stomach for analysis had not always proven easy. A variety of investigative methods had been attempted including those of René Réaumur who had trained a pet kite to swallow and regurgitate food-filled tubes in order to show that digestion was primarily a process of chemical dissolution rather than trituration and putrefaction.[2] Italian physiologist Lazzaro Spallanzani, meanwhile, had established the solvent powers of saliva by regurgitating linen bags. He was also in the habit of inserting putrid flesh into the stomachs of living dogs as part of an attempt to ascertain whether the powers of the organ's juices could make the flesh fresh again.[3] Yet during the nineteenth century, investigations took on distinctively new forms than previously. Laboratory experimentation increasingly aspired to grounding understandings of illness and disease firmly in new regimes of modern science. Experimental physiology became far more intense and continuous than in earlier periods and was almost entirely pursued in institutional settings.[4]

Given the stomach's centrality to nineteenth-century British medical activity, it might be expected that its transformation into an object of sophisticated medico-scientific inquiry would have been greeted enthusiastically, especially given physiology's fundamental claims to offer improvement in the clinical setting. As suggested above, despite the high levels of attention directed towards gastric complaints, knowledge of the organ had always remained vague from a medico-scientific point of view. Physicians had therefore always struggled with a limited ability to discriminate between even minor cases of indigestion and more serious complaints such as gastric ulcer or even cancer of the stomach. The diagnostic symptoms of these were not always obvious, consistent or easily distinguishable from one another.[5] Despite the attempts of early nineteenth-century pathological anatomists to elucidate new knowledge, even basic matters remained remarkably unclear, including what shape the organ was in its living condition, or even if there was such a thing as a standard-shaped stomach. It is unsurprising to find that the digestive system and its component organs came under the close scrutiny of physiological enquiry.

A crucial element of the agenda of physiologists was the hope that clinical practice would become increasingly complemented by discoveries and technologies accrued from laboratory enquiry.[6] Yet although this aspiration appeared worthy in theory, especially given its inherent claims for the relief of patient suffering, the incorporation of laboratory medicine into clinical practice was not fully implemented in Britain.[7] Shifting conceptions of the activities of the medical profession were central to this. Resistance partly resulted from tensions that emerged between traditional and modern medical ideologies. British systems of medical education had, for the most part, placed less emphasis on scientific research than their European counterparts.[8] The desire of the cultured, gentlemanly physician to retain his social prestige by stubbornly continuing to employ the traditional methods embedded within his general education was an important factor that contributed to a lack of enthusiasm for engagement with new practices. The British clinician intended to maintain perceptions of his work as an intuitive, clinical art rather than a technological, scientific practice, insisting that new diagnostic procedures threatened to make his art overly mechanical. Medical practice, it was feared, was coming to resemble the work of the artisan rather than that of a refined professional gentleman.[9]

The behaviour of British physicians in this period can also be understood by looking beyond a simplistic science versus intuition dichotomy. Laboratory medicine became tainted by the spectre of its association with animal experimentation. The image of the modern medical professional as vivisectionist was culturally pervasive, and provided particularly negative depictions of him as an undertaker, or at best advocate, of cruel acts of medico-scientific torture on living animals.[10] Gastric technologies, especially the stomach tube, became emblematic symbols of the tensions in existence between the discipline and those appalled by the nature of experiments on animals and its implications for the future of medical practice. This also impacted upon the interaction between the various medical disciplines.

The Laboratory and the Stomach

Physiology gained increasing influence within Western medicine during the late nineteenth century. Within European and American medical schools, workspaces were constructed for chemists who were dedicated to experimenting with and examining bodily secretions. The rise of the field was remarkably rapid and by the 1870s physiology was firmly consolidated as a discipline with specialized university chairs and laboratories in existence in various Western countries. Digestion became remarkably central to physiological enquiry. The influential German chemist Justus von Liebig (1803–73), for instance, expressed a strong interest in scientifically determining the nature of the digestive processes. By the 1840s he had analysed food digested by pancreatic juice, and proven that it contained carbohydrates, fat and protein.[11] In France, meanwhile, the leading physiologist Claude Bernard (1813–78) produced major works on important internal phenomena including pancreatic function.[12] Yet, Bernard had a far wider agenda for his discipline. He argued that the clinical approach was proving to be far too limited to continue to act as the foundation of medicine, and insisted instead that it was physiological investigation which held the potential to reveal important insights denied to the clinician. For this reason, he thought that the laboratory should be awarded a superior status to the bedside as a source of bodily knowledge.[13] To accomplish this aim, Bernard insisted that it was necessary to show that medico-scientific research held practical clinical potential, and that it was able to offer practical solutions to a variety of medical problems. It was therefore within the interaction between physician and patient that the authority of experimental physiology was put to the test.[14]

The physiology of digestion certainly became more clearly elucidated as a result of laboratory medicine, for instance. The influence of food and drink on pepsin levels, and rates of absorption in the stomach, were successfully investigated. This represents further changes in the way in which medical scientists came to adopt a more scientific approach to the matter of food. The impact of differing levels of acid within the stomach upon the speed of digestion was also discerned, while attention began to shift from the digestive processes of the stomach towards its workings in the smaller abdominal organs. The liver, for instance, became recognized as a blood refiner which separated carbonaceous matters, while the pancreas became described as a salivary gland containing high levels of alkali. This had a visible impact on clinical diagnosis. The term 'acid dyspepsia' came into more common usage, a reference to the heightened knowledge available on the substances contained within the gastric juices. The source of this acid was now believed to be secretions from the tubular glands, a notion that helped firmly to replace long-standing views regarding the role of fermentation in producing gastric chemicals.[15] Medical authors attempted to include more stringent varieties of dyspepsia even in common-sense texts on how to manage the stomach. These included new diagnostic categories such as plethoric dyspepsia, anaemic dyspepsia, hepatic dyspepsia and renal dyspepsia.[16] Of course, this is not to say that these concepts were universally accepted. Homeopathic writers, for instance, continued to employ terms such as tobacco, tea, tight-lacing and even London dyspepsia towards the close of the nineteenth century.[17]

Overall, however, the chemical analysis of gastric contents was presented as a tool which could be shared by laboratory workers and physicians alike. Physiologists were eager to show that by obtaining and scrutinizing samples of human gastric acids, the physician would be placed in a better position to make more accurate distinctions between the various gastric illnesses, with the ultimate result of this being the recommendation of appropriate forms of treatment for the suffering patient's ailments. Once retrieved, it was suggested that gastric contents could easily be filtered and analysed in order to determine diagnostically useful factors, such as levels of acid or peptic power.[18] Particularly high levels of hydrochloric acid discovered in a patient, for instance, became hypothesized as being symptomatic of gastric ulcer disease. Furthermore, laboratory methods held claims to be able to yield data which, due to their scientific rather than intuitive nature, would inherently hold more precision than the diagnosis of the physician.[19]

The stomach began to be accessed with a wide range of innovations in this period, many of which are now long forgotten. These technologies were emblematic of the new aspirations of the discipline of physiology. New York physician Max Einhorn (1862–1953) showed particular enthusiasm in inventing such technologies.[20] One of his creations was the 'stomach bucket', later developed into the 'duodenal bucket', which was a small oval-shaped container with an opening attached to a silk thread. This would be swallowed by the patient and quickly reached the stomach, allowing for gastric chemicals to be collected and drawn back out of the body for scientific analysis.[21] Einhorn also developed the gastrograph, which consisted of a stomach tube with a hollow platinum ball attached. As the stomach moved naturally, the ball came into contact with an electrical current which allowed gastric movements to be recorded on an external apparatus.[22] Similarly, George Herschell, senior physician at Queen's Jubilee Hospital, London, pressed for the adoption of the gastrodiaphane within general practice, an instrument consisting of a soft rubber tube fitted with a bulb, intended to illuminate the stomach.[23] A further therapeutic method developed by Herschell was auto-lavage. This involved the use of a stomach tube connected to further tubes all attached to a metal douche can filled with a cleansing solution and a receiving pail. The can would hold at least two litres of liquid, and was hung on a nail in the wall above the patient's head whilst seated.[24]

New techniques involving filling the stomach with liquids or gases were also developed with the aim of being able to improve clinical diagnosis. These were intended to manipulate the size and shape of the stomach to render its medical conditions more visible to the practitioner. The German physician Franz Riegel (1843–1904) introduced the 'Clapotement', or 'Splashing Sound' device. While experimenting, he had discovered that if a patient had digested large quantities of fluid then a sound could be heard if the fingertips were pressed into specific areas of the abdomen. If a sound could be heard below the normal lower limit of the stomach, then this might indicate dilatation of the stomach or displacement. If the 'splashing sounds' could be heard between four and six hours after a meal, then delayed emptying was thought to be occurring, a symptom considered to be closely associated with atony of the stomach. Meanwhile, localized sounds around or below the umbilicus appeared indicative of gastroptosis.[25] An innovation which inflated the stomach with gas consisted of a double-rubber bulb that could gauge and control air quantities inside the organ. It was hypothesized that a relaxed and dilated stomach would require significantly larger quantities of gastric juices than a healthy organ with normal muscular tone.

Laboratory medicine also encouraged the development of therapeutic technologies for the gastric system. Electricity, for instance, became recommended for therapeutic application following experimental enquiries into its medical value. Einhorn constructed an electrode that was similar to the stomach bucket, but which contained an additional rubber tube connected to an external battery. This would be placed directly onto the skin immediately above the stomach, or one might be placed within the stomach with another being placed on the skin in a corresponding area above. Faradization was argued to hold therapeutic benefit for atony of the stomach wall, dilatation of the stomach, gastralgia and hypochlorhydria.[26] A further therapeutic innovation derived from the laboratory treatment was the elimination of excess acid problems by the application of alkali substances which were intended to saturate acids. It is worth noting, however, that many of these antacids were seen to be more palliative than curative. Fears were raised that the temporary relief offered by alkalis might lead to patients becoming addicted to consuming the substances. Their long-term usage also seemed to risk the production of long-standing digestive problems.[27]

The stomach tube was the most important, and most enduring, investigative innovation that stemmed from physiology. It was introduced in 1868 by the German physician Adolf Kussmaul (1822–1902), who perfected it by experimenting on a professional sword-swallower. The design of the tube facilitated the emptying of the stomach so that its contents could be chemically analysed.[28] The German gastroenterologist Carol Anton Ewald (1845–1915) further developed this method of intubation, with his main contribution, whilst working with his colleague Ismar Isidor Boas (1858–1938), being the introduction of a standard test meal to improve the accuracy of gastric analysis. Specific foods would be given to a patient, with the retrieval of stomach contents taking place at a scheduled interval afterwards, a method thought to lead to more uniform results.[29] Ewald's test breakfast contained bread and weak tea which was to be digested an hour before the extraction of stomach contents. However, other variations included combinations of minced meat, tea, toast, mint, soup, scraped beef and wheaten bread.[30]

Overall, the production of these innovations is indicative of the intensive fascination that the digestive system, and the stomach, held for late nineteenth-century medical scientists. It is also evidence of an evocative scientific influence on the management of stomach complaints that brought with it a new range of analytical tools and technologies. These also served professional purposes. The ability of physiologists to infiltrate the stomach acted as a marker of the burgeoning prowess of laboratory science in mastering internal illness, just as it had done for pathological anatomists.

The Stomach and the Laboratory Animal

From the late nineteenth century onwards, potential clearly existed for the management of gastric disorders to become directed by a new ethos of scientific accuracy and technological investigation. How, then, can antagonism and wariness towards these new methods be accounted for given physiology's claims to be able to resolve the long-standing difficulties associated with the management of gastric illness? In many ways, this is an especially surprising scenario in Britain given the existence of a pronounced cultural anxiety over gastric conditions such as dyspepsia. Suggestions that low levels of material support for research thwarted the scientific endeavours of British physiologists are problematic in some respects as they do not fully explain the rejection of technologies and procedures at the clinical level. After all, these could easily have been imported and incorporated into teaching and practice.[31] Similarly, claims that new diagnostic technologies were rejected because of their personification as a threat to an intuitive clinical art which medical practitioners were anxious to defend is also limited.[32]

It is necessary at this point to look briefly at wider cultural discussions regarding what it meant to be 'scientific' in the laboratory before returning to the subject of the stomach. Advocates of laboratory medicine regularly met scepticism, apathy, indifference and suspicion in Britain, and this opposition extended far outside of the clinical sphere. Much of this hostility resulted from its direct association with animal experimentation, a practice that proved inseparable from laboratory research much to the detriment of the discipline. Strong feeling towards experimental procedures existed both amongst members of the general public and within certain sections of the medical profession in Britain.[33] To adopt a scientific approach to medicine in this period therefore also meant to risk association with a wide variety of negative representations. Notably, a particularly prominent image persistently forwarded was that of the modern medical professional as a brutal, dispassionate torturer rather than caring healer. For many critics, medicine appeared to be becoming far less humane than it had traditionally claimed to be, and now seemed to serve science rather than mankind. While the art of medicine had been interested primarily in restoring sick persons to health, opponents insisted that it was becoming overly concerned with discerning the laws which regulated human life, in both its healthy and diseased states, rather than in providing cures.[34]

The inclusion of scientific principles into medicine was therefore problematic as it prompted concern that clinicians increasingly perceived their public responsibilities as normalizing a deviant physiological process rather than caring for a sick human being.[35] New investigative methods designed to explore the gastrointestinal tract were therefore inherently wrought with problems of internal ethics and public accountability. These were heightened by suggestions that by the turn of the century the clinical benefits of physiology, as distinct from those of bacteriology, pharmacology and pathology, appeared to be few indeed. This made ethical questions regarding connections between suffering and therapy particularly challenging for the laboratory scientist and for the clinician who might make use of new gastric technologies.[36] What was at stake in the British vivisection controversies was not only the moral character of experimental physiologists, but also the image of medical professionals more generally.[37] New relationships therefore had to be negotiated between science and medical practice, which took into account themes of professional identity and moral legitimacy.[38] Overall, antivivisection sentiment, including that from within the medical profession, exerted a powerful inhibitory influence on the adoption of the ideals and technologies of physiological exploration, more so in the British clinic than in other countries.[39]

Given the centrality of investigations into the digestive system to the emergent discipline of laboratory medicine, it is unsurprising that experiments undertaken into the workings of the gastric tract were widely utilized within antivivisectionist propaganda. The procedures and technologies newly associated with gastric illness came to act as an emblematic marker of medico-scientific brutality and provided a pertinent symbol of the controversial new direction that medical science was pursuing. Antivivisectionists proudly attempted to make the more dubious aspects of laboratory research into the digestive system publicly known. Any claims that medico-scientific work was leading to significant curative benefits in this area were strongly refuted, or at best portrayed as disproportionate to the high levels of suffering imagined to have been instigated during the complex processes of knowledge accruement. Tellingly, the University College London physiologist Ernest Henry Starling (1866–1927) was particularly eager to show that his research had greatly enhanced knowledge of digestion, and that his investigations into the stomachs of living animals had provided far more accurate knowledge than almost a century of investigations into human morbid anatomy.[40] Yet, Starling's work proved objectionable on two levels: firstly, his use of mutilating techniques upon laboratory dogs; and, secondly, the extent to which he was confidently able to claim that he was developing information intended for medical, rather than scientific, interest.

Whilst experimenting, Starling gained access to the stomachs of living laboratory dogs by creating fistulous openings, a process which involved forging a hole into the body of the animal through which its stomach could be accessed and samples of digestive juices collected with ease. However, this technique had limitations due to the difficulties in evoking gastric secretion by physical stimulation of the exposed stomach. The amount of gastric acid produced via these techniques was therefore relatively small. Starling went on to employ alternative procedures to work his way around this problem, yet by doing so laid himself open to claims of cruelty and needless bodily mutilation. In order to collect higher levels of gastric chemicals, he cut the oesophagus of the laboratory dog's neck in addition to creating an abdominal fistula. By these means, it was still possible to feed the animal by pouring liquid food through the oesophageal opening, meaning that the animal could stay alive whilst being experimented upon.[41]

It was only by encouraging the dog to eat orally, however, as it would do normally, that the high levels of gastric juices requisite for analysis would be produced. The dog could still take food into its mouth, chew it and swallow it yet, much to the dog's surprise, the food would always fall out of the hole made in its neck. If the laboratory dog was provided with food when in a state of extreme hunger, it might avidly attempt to eat in this futile manner for hours, without ever realizing that food would never reach its stomach. An alternative procedure described by Starling involved an experiment where a dog was continually excited for up to ninety minutes by being shown meat in order to stimulate the production of gastric juices. The dog's stomach would continue to produce liquid as long as it continued in its attempts to eat. Such techniques became known as 'sham feeding' and rendered the dog as a suitable object from which digestive juices could be collected for scientific analysis.[42]

Such ethically questionable procedures required justification to both the public and opponents within the medical profession, with the test of their value being their obvious clinical worth. The accruement of scientific knowledge as an end in itself was far from enough to justify the infliction of pain onto a living animal. Certainly, Starling's investigations were easily construed as a series of cruel, mutilating and torturous experiments which held little, if any, transferable value for the human patient. Starling claimed, for instance, that by utilizing sham feeding, he had ascertained that

> the afferent channels for this reflex may be therefore either the afferent nerves from the mouth, or, when the *idea* of food is involved, any of the nerves of special sense, such as sight, smell or hearing, through which these ideas are called forth.

Yet this was noted to have failed to lead to any discernible improvement which could be transferred into medical practice.[43]

Within their propaganda, antivivisectionists fiercely complained that such conclusions were of scientific interest only, containing no transferable medical value whatsoever. Starling's work was vigorously condemned in publications including the popular antivivisectionist magazine the *Animal's Defender and Zoophilist*, which argued in 1915 that despite devoting months of research to the study of pancreatic diabetes, Starling had only got as far as concluding that 'it is still very difficult to say definitely why the removal of the pancreas brings about this condition, or what disturbance of metabolism is primarily responsible for it'.[44] Again, this finding alone was not enough to convince sceptics of the practical benefits of scientific medicine.

Sham feeding emerged as a persistent target of the antivivisectionist movement, and became one of the prime symbols of the medical brutality alleged to have permeated the discipline of physiology. Its members persistently provided grotesque and evocative accounts of such experiments and their appliance for purposes of scientific curiosity alone. One story reproduced in the antivivisectionist publication *The Shambles of Science*, entitled 'Fun', recalled a visit to a physiology lecture. The anonymous author claimed that:

> The lecturer describes certain experiments on dogs amid the laughter of the audience. The oesophagus has been cut and a fistula established, so that the food taken fell down on the floor instead of passing into the stomach. The dogs ate and ate and ate – they were frightfully hungry – and were much surprised to see the food fall out; they tried again with the same result. They could go on like that for hours! How comical! How clever of the physiologist who tried this! Aren't animals stupid? During the process of eating, the stomachs of the dog secreted gastric juice. This is an instance of 'psychic secretion'. Awfully interesting! Marvellously clever!

The author went on to state:

> When describing some other experiments, where food has been introduced directly into the stomachs by means of fistulae when the animals had been asleep, the lecturer jokingly said that this was rather difficult, for you had to be careful not to awake the dogs, but only their stomachs.[45]

Charles Bolton (1870–1947), director of pathological studies and research at London University College Hospital, was a further physiologist attacked by antivivisectionists. In his writing, he professed that his employment of a scientific approach to medicine had done much to advance clinical diagnosis, and in particular with the clinical management of ulcer of the stomach. Furthermore, he insisted that gastric disease could only be understood and cured if the physiological processes of the stomach were fully elucidated. Like Starling, he maintained that it was not possible to observe accurately the workings of the digestive system in a healthy, living human, and that it was essential to align the findings of morbid anatomists with the results of clinical research.[46] Bolton also experimented on a range of animals including dogs, monkeys and cats. Yet when he published his research, his techniques were reprinted, and fiercely condemned, in the antivivisectionist press.

The movement's authors raised a variety of ethical issues surrounding Bolton's work on ulcer of the stomach as he had attempted to produce the disease artificially by injecting doses of acid into the stomachs of monkeys, working on understandings that hydrochloric acid was the principal cause of the complaint. The *Animal's Defender and Zoophilist* claimed that Bolton made no mention of whether or not he had undertaken his experiments using anaesthesia, and concluded that he probably had not. However, even if anaesthesia had in fact been administered, the torture inflicted by the resultant ulcer, combined with the irritant effects of acid poisoning, would still have proven highly traumatic for the animals. The author therefore concluded that:

> It cannot be pretended that these cruel experiments were performed for the benefits of its victims, nor can it be argued that because we can burn holes in their stomachs with irritants and destructive acids we can advance our curative methods for a well-known and cruel disease. The experiments were undertaken for purely scientific reasons, and the whole business was another proof that research of this kind is merely cruelty to animals that should be suppressed by the law.[47]

Antivivisectionist literature continued to detail a wide range of experiments made upon the laboratory animal, eagerly reprinting and reassessing physiological reports from medical journals and textbooks. These were described as the tearing away of the stomachs of various dogs and replacing them with a pig's bladder to produce artificial vomiting; the continuous injection of alcohol every half-hour into a dog's stomach until it eventually died; and the opening of the abdominal organs of two rabbits and subsequent transfer of the contents of one into the other to investigate the physiology of pregnancy.[48] Other alleged procedures include the uniting of the bowel to the stomach and the cutting open of the stomach of a dog in order to insert the ear of a live rabbit until it was eaten away by the animal's gastric juices.[49]

Clearly, scientific investigation into the stomach contained ethically challenging implications. Given the heightened importance of the organ in the

British context, and the necessity of accruing clinically applicable interpretations of gastric illness, it is unsurprising that investigations into the organ became readily transformed into one of the most symbolic and pertinent examples of claims of modern medical cruelty. The treatment of dyspepsia and gastric ulcers would, if successfully achieved, have convinced many of the validity of physiological intervention. Pointing out its failure to do this in an era characterized by a heightened awareness of the individual and collective implications of widespread gastric conditions emotively strengthened the claims of both antivivisectionists and conservative elements of the medical profession.

The Physician's Negotiation with Scientific Medicine

Scientific medicine in this period clearly held far broader meanings than its preferred positive association with the progression of knowledge. It can be observed carrying darker connotations and implications. The incorporation of new methods of gastric analysis at the clinical level necessitated the utilization of new, intrusive forms of scientific technology which held multifaceted meanings for both patient and physician. The controversies surrounding the relationship between technology and pain played a prominent role in the opposition to laboratory medicine that emerged within elements of the medical profession. It also played a notable role in patient perceptions of the clinical experience and amongst those physicians who may well have been open to the idea of new forms of gastric analysis.

Anxiety regarding experimental medicine was not new. In fact, the British public had a long-standing tradition of being suspicious of the medical professional's seemingly eager desire to perform experiments on their bodies. Concern over this had been accelerated by the early nineteenth-century bodysnatching panics which had resulted in the passing of the Anatomy Act of 1832. This legislation had made the body of the workhouse pauper available for medical research should no one claim it.[50] Against this historic backdrop, claims that the gaze of the medical professional was now turning towards experimentation on the *living* human body proved to be an effective rhetoric for those actively opposing the techniques of the laboratory. Fears of human experimentation were clearly explicit within the antivivisectionist representations and it appeared perfectly plausible to many contemporaries that the human too might fall victim to the cruel urges of the modern medical professional if the ethos of laboratory science was allowed to pervade too far into the clinical experience.[51] The patient, it was feared, might become an object and target of scientific inquiry or brutality. To modify laboratory instruments for the human inevitably meant, in this period, to dissociate treatment from any negative associations held with medical science.[52]

Discernibly embedded deep within the clinical use of such technologies in Britain were questions related to the problematic relationship between torture and cure. In the case of gastric analysis, laboratory techniques designed to analyse the stomach appear to have been applied in a restrained way dependent on the extent to which they could justify their use as being of clinical value. It was necessary to maintain the impression that the application of invasive and uncomfortable procedures was in fact worthwhile and produced significantly improved levels of diagnosis and cure, if dissociation from vivisection controversies was to be fully achieved.

Discussion of the chemical test provides one example of this. Supporters of modern scientific medicine advocated new methods such as the vivid-green salt test, a procedure that involved adding hydrochloric acid to a solution of extracted gastric contents, as they appeared particularly simple to use in the clinical setting. Claims were forwarded that scientifically accurate conclusions might be reached about a variety of abdominal illnesses by the employment of these tests which were significantly superior to diagnostic results which relied upon traditional methods of observation. However, in 1899, the *British Medical Journal* suggested that these apparently simple tests in fact produced findings that were no more accurate than those obtained by traditional methods of examination. Gastric substances including peptones and neutral salts, it was suggested, interfered with chemical reactions. Furthermore, those methods appeared to be too complicated for practical, clinical use. The journal concluded that 'at present, indeed, a ready method, suitable in clinical practice of the detection of free hydrochloric acid in organic liquids is a desideratum'.[53] The results of such tests were also persistently criticized for not being uniform enough to be able to reach decisive diagnostic conclusions about gastric disease. It appeared, for instance, that vomit in stomach cancer regularly contained small amounts of blood, which was also recognized as a symptom of gastric ulcer disease, meaning that diagnostic distinctions could not always be easily made. Accordingly, physicians often resorted to more familiar methods, claiming that the true test of determining malignant stomach disease from ulcers was to determine the presence or absence of a tumour through the traditional method of undertaking a physical examination of pain.[54]

In this case, this does not suggest that chemical analysis was by any means rejected. In fact, physicians appear to have utilized them initially but soon found them to be unsatisfactory. Instead, it seemed to be the lack of visible improvements in diagnosis that proved to be the decisive factor in the physician's choice not to use them. Overall, chemical tests would be rejected if their results appeared disproportionate to their diagnostic or therapeutic value. Yet by rejecting such technologies at the clinical level, physicians were also self-consciously dissociating themselves from those aspects of laboratory science which seemed to produce little of curative value. The apparent complexity of new, scientific methods, and the extra time spent subjecting the patients to their associated gastric technologies, were clearly judged in relation to their diagnostic value. To obtain appropriate levels of accuracy, it often appeared necessary to employ methods that were so complicated, difficult and time consuming that they were out of the reach of anyone unskilled in chemical analysis, or who did not have access to the large amount of apparatus often necessary. Busy practitioners fiercely complained that they simply did not have enough time on their hands to deal with such lengthy processes or to employ them in an accurate, conscientious manner.[55] Complexity combined with impracticality and relatively little diagnostic or therapeutic yield proved highly problematic in an environment where pressure to express the clinical worth of new procedures of gastric analysis was high.

Yet, perhaps most significantly of all given the problematic questions regarding pain and the laboratory, patient discomfort was the leading argument employed by physicians when justifying their perseverance in utilizing familiar diagnostic and therapeutic methods. It is unsurprising that discussion of the implications of laboratory medicine focused so intensely upon the stomach. It was an organ positioned centrally within the body, which rendered it especially inaccessible in comparison to many of the other major organs. Innovations such as the stomach tube were therefore amongst the most intrusive of late nineteenth-century technologies as their use necessitated entry into the deepest recesses of the inner body. Of course, it was for this very reason that their successful adoption in the clinical setting, if fully accomplished, would have helped to prove the prowess of modern medical science. The stomach itself came to act as the test of physiology's ability to transfer its procedures successfully into the clinical setting.

The gastroscope, for instance, consisted of a rigid, bent metal tube containing a prism placed at an angle and an electric lamp at its nosepiece. The device required the heating of the lamp whilst inside the stomach and necessitated a constant stream of water flowing through two water channels to reduce its temperature. Its application on the patient was somewhat impractical, with it being necessary to administer large doses of morphine. Such deep anaesthetization meant that the patient's throat reflexes were stopped. To administer the procedure, the patient then needed to lie on his back with his head over the end of a table, or be uncomfortably positioned propped up on a chair with his head tilted backwards. This was described in one account as 'an exquisite surgical manipulation which should only be undertaken by the most expert', meaning that the device was of limited practical clinical use.[56]

Patient discomfort persistently tainted discussion of gastric technologies associated with the laboratory. When coupled with implications of low diagnostic yield, a convincing argument was able to be formulated against the intrusion of physiological science into the doctor–patient encounter. According to one medical author, the enthusiasm of many of those who had initially been eager to apply gastroscopic methods had been quickly dampened by accidental, sometimes fatal, perforations of the gullet or stomach. The gastroscope, according to his account, was proving more dangerous as a diagnostic tool than exploratory laparotomy in the hands of a competent surgeon not least because it was alleged to involve the 'blind insertion of an angular, rigid rod through the gullet into the stomach'.[57]

Scientific methods involving the expansion of the stomach with water can also be found being condemned as inaccurate, unpleasant and highly dangerous should an anomaly such as an ulcer be present. Procedures involving listening for sounds were dismissed as too complicated for practical use, as were devices intended to dilate the stomach with gases which were feared to have potentially damaging effects on the patient's health. This risk did not appear to be proportionate to the negligible improvement in diagnostic accuracy.[58] Similarly, inflating the stomach with gas was a procedure criticized on the grounds that the patient was likely to gag whilst it was being undertaken, and also, more importantly, that many patients would object to having the physician's breath blown into their stomach.[59] Often, then, it was the intrusive nature of emergent technologies which physicians cited when detailing their reasons for their rejection of laboratory techniques.

Whether or not these authors exaggerated their experiences with new forms of gastric analysis in order to confirm pre-existing views on the validity of their traditional method, or if they provided an accurate account of patient discomfort is unclear. Yet what is important is their evocative usage of association with pain, a theme closely associated with the problematic aspects of physiological enquiry. This association implies a deeper awareness of the connotations surrounding laboratory medicine and an attentiveness to the likely sources of patient wariness which would have impacted upon perceptions of their professionality amongst the public.

The stomach tube became central to these strands of professional discourse. In the laboratory setting, the use of the tube had ascertained that the secretions of the parotid and sublingual glands had the consistency and colour of water, whilst the syrup-like fluid of the submaxillary gland had more solid constituents. The effects of different substances and stimulants in raising secretion levels which these glands produced were also more clearly elucidated.[60] The stomach tube clearly held some degree of scientific value. Yet the ultimate test of its usefulness was in the clinic. However, within this setting it was not unheard of for accidents to occur such as the unintentional swallowing of the device which would then sink into the patient's stomach where it could never be retrieved. Even when performed successfully, the comfort of the procedure was dubious. It was necessary, when inserting the tube, to make sure that mucus and saliva did not drop from the patient's mouth and for there to be receptacles on hand in case of vomiting.[61] It was also frequently found that if the stomach contained food residue, then as soon as the tube was introduced some of this would be vomited all over the hands of the physician.[62]

Furthermore, intrinsic design problems were claimed to be common in the tubes available to British practitioners, which became criticized for being designed without careful consideration of patient comfort and the reduction of internal pain. Many practitioners used tubes containing eyes made by directly punching a hole in the rubber, a method thought to have created sharp cutting edges which acted as curettes once inside the body. This risked injuring the mucous membrane of the patient's stomach. If the tube was to be suddenly removed, the sharp edges of the eyes might cut the stomach wall, and even detach linings of the membrane. It was also noted that it was easy for the tube to become blocked by food, rendering it diagnostically useless. Finally, many practitioners mistakenly believed that small tubes could be inserted with more ease than those of a larger size. However, in reality, larger tubes could be grasped and swallowed by the oesophagus more comfortably, being less liable to bend or kink when applied.[63]

Patient opinions towards the tube were also claimed to have been mostly negative. Tellingly, George M. Niles, author of *The Diagnosis and Treatment of Digestive Diseases* (1914), suggested that public attitudes towards the stomach tube could be described as 'actual repugnance', claiming that:

> I have known many patients, who have spent sleepless nights in awesome anticipation of the trying ordeal, and others, who would suffer for months, rather than submit to what they considered a horrible torture. So often do I hear an intelligent patient say – 'Doctor, I would have been to you for aid long ago, had I not dreaded to take that awful stomach-tube'.[64]

The attitude of the patient was certainly perceived as a stumbling block in the adoption of the tube in clinical practice. George Herschell claimed that methods of gastric analysis were not employed in Britain with the same levels of enthusiasm as on the Continent due to a deeply rooted idea amongst medical men that the patient would never submit to such forms of diagnosis and treatment. Accordingly, he provided the medical profession, in his advice on intragastric technique, with lengthy instructions on how to improve the tube's reputation amongst patients, starting with tips given on how to introduce it into the body painlessly. For him, it was the nervousness of the patient which would hinder the success of the investigation, rather than clinical inability.[65] Herschell insisted that in order to incorporate scientific methods into clinical practice, it was necessary to distance modern technologies from such negative representations that proved persistent in Britain throughout the period in question.

The Suffragette's Encounter with the Stomach Tube

Fears of the cruelty and pain of the laboratory being transferred directly into the clinical setting appeared to be turning into reality during the controversy surrounding the suffragette hunger strike in British prisons from July 1909. The stomach tube became utilized as a symbol of political and sexual oppression following its application to forcibly feed imprisoned militant suffragettes. It came to act as the ultimate emblem of medical brutality, exemplifying all that was perceived to be wrong with the scientific direction that modern medicine was pursuing and the perverse uses to which it might be put. Representations of the stomach tube self-consciously projected images of patients at risk from the cruel gaze of the modern medical professional. Furthermore, militant suffragettes were well aware of the wider associations which the technology held with dubious forms of medico-scientific enquiry.

The application of the stomach tube for the purpose of feeding patients who were, for whatever reason, unable to eat, had been a standard medical procedure throughout the late nineteenth century. Forcible feeding was utilized in asylums to save the lives of patients who were considered to be so severely ill that they had stopped eating or had become unable to swallow food.[66] The procedure was generally depicted as safe, with only minor complications arising such as nausea, vomiting, stomach cramps and diarrhoea. Furthermore, asylum doctors claimed to have become very skilled at this procedure, quickly learning at what speed and temperature food could be administered without causing discomfort.[67]

When applied on a patient who was not ill, however, but who might have voluntarily chosen to stop eating, the procedure became ethically problematic. This had been pointed out occasionally. An article had been published in the *Lancet* in 1872 following the use of forcible feeding on a murderer who had been imprisoned in an asylum. The author claimed that 'if anyone were to ask me the worst possible treatment for suicidal starvation, I should say unhesitatingly, "forcible feeding by means of the stomach-pump"'. He went on to claim that forcible-feeding by the mouth ought to be abolished forever as it was more concerned with violence and force than cure.[68] In a further article, he described the stomach pump as 'the most unmerciful engine for the purpose of feeding that has ever been invented'.[69] This article, however, seemed to provoke little response and use of the technology remained relatively inconspicuous throughout the rest of the century. Yet the tube came to carry negative connotations amongst the public. In cases of anorexia nervosa, families would avoid incarceration in an asylum as forcible feeding was perceived to be an overly harsh punishment designed to normalize deviant female behaviour.[70]

When the British government made the decision nearly forty years later to utilize both the stomach tube and stomach pump to forcibly feed suffragettes undertaking hunger strikes, a public controversy emerged. In July 1909, Marion Dunlop (1864–1942) was imprisoned for an act of suffragette militancy and refused to carry on eating unless she was awarded political prisoner status. Partly out of fear that she might die and become a martyr for her cause, the British government made the decision to release her after 91 hours of self-starvation. However, organized groups of imprisoned suffragettes, spurred on by the effectiveness of Dunlop's activities, quickly adopted the same strategy. Unwilling to release all of them, a decision was made by the Home Secretary in September 1909 for medical officers to force the prisoners to eat against their will.[71] Although a public and medical controversy flared up surrounding the ethical dimensions of non-consensual feeding, prison medical officers were legally obliged to comply.[72] Notably, the British scriptwriter Kitty Marion (1871–1944), imprisoned for throwing stones at a post office in Newcastle, claimed to have endured 232 forcible feedings in prison whilst on hunger strike.[73]

The seemingly unnecessary use of the tube became immortalized in some of the most provocative images of the Edwardian period. It also became challenged in various public displays of protest that encouraged the ethical questions surrounding the use of new medical technologies to become heavily debated. Members of the medical profession became directly implicated in alleged scenarios of brutal torture with the stomach tube. In November 1909, protestors smashed the windows of the house of Dr Cassell, the deputy medical officer at Winson Green Prison, who had assisted a prison doctor in forcibly feeding suffragette prisoners on this basis.[74] At a speech at the Child Study Society on the subject of 'The Child Criminal', given a fortnight later in London, Dr W. C. Sullivan, medical officer of Holloway Prison, was repeatedly interrupted by suffragettes, who asked him how he could address such an audience when he took part in the forcible feeding of women in Holloway Prison. They went on to describe his work as 'dirty' and 'absolutely degrading'.[75] Prime Minister David Lloyd George (1863–1945), meanwhile, also faced protests when at the Savoy Theatre that month involving campaigners carrying flags bearing various phrases including 'Votes for Women' and 'No Stomach Tubes'. Although the women were escorted off the premises, in the second act of the play two more protestors stood up in the stalls holding up feeding tubes and banners inscribed with the phrase 'No Forcible Feeding'.[76] In these instances, the stomach tube was held up in highly public displays of protest as an exemplar object of medical brutality.

The association of the instrument with the ethical questions surrounding laboratory medicine was not lost on the militant suffragists. Links between antivivisectionism and feminism were often strong, and this heavily influenced the nature of suffragist propaganda against forcible feeding. Caring women with feminist tendencies allied themselves with the laboratory animal not least because this offered them a further platform in which to oppose male-orientated militaristic and inhumane expressions of modern science.[77] By the early 1900s, many British people were convinced that the suffragists and the antivivisectionists were in fact one and the same. It seems plausible that many women perceived the tortured animals as being victimized by the very same dominant forms of male-orientated scientific ideology as themselves. The circumstances of the animal reflected how they felt about their own social condition and entrapment.

The evocative image of a painful, degrading and unnecessary procedure involving varying processes of internal bodily manipulation closely resembled culturally pervasive ideas of the experience of the laboratory animal. Implications that the suffragettes were being treated in a similar fashion to the laboratory animal formed an integral part of their representations of the male authority of the medical profession. In December 1913, a large demonstration was held against forcible feeding at Queen's Hall, London. At this, the Bishop of Kensington stated that to inflict useless pain via forcible feeding constituted torture. He employed the example that if someone was caught flogging a horse, then the Society for the Prevention of Cruelty to Animals would be informed, on the grounds that it was inhumane and cruel as well as useless.[78] Constance Lytton (1869–1923) also reminded those reading her accounts of the similarity of her position to that of the laboratory animal when she described her 'feeling of complete helplessness, as of an animal in a trap, when the operators come into one's cell and set to work'.[79] Furthermore, suffragette authors claimed that the prison doctors and wardresses described their duties as 'feeding them [the prisoners] like chickens'.[80] Similarly, Dorothy Pethick, a prisoner at Newcastle, complained that she felt as though she was being treated like 'a piece of cattle'.[81]

Medical duplicity formed an integral part of suffrage portrayals of forcible feeding, neatly mirroring the claims made by antivivisectionists regarding the dishonesty of physiologists. Deeply embedded within the evocative descriptions of the prisoner's experiences of the stomach tube were claims that unnecessary, brutal and painful forms of medical torture with no curative benefit were taking place. Prison medical officers insisted, on the contrary, that they were saving the lives of female prisoners who would otherwise have died through starvation or who would have had at least begun to suffer ill health, allowing them to claim that they were acting compassionately. However, the prisoners responded to this by pointing out that they were far from being in a state of ill health when the stomach tube was applied. Nor were they on the brink of starvation, so they insisted. If anything, prison medical officers were claimed to have paid little regard to the well-being of the prisoners, instigating procedures which ultimately created more health problems than they had intended to resolve.[82]

This allowed the militant suffragists to portray forcible feeding as a procedure deceptively masquerading under false claims of medical benefit. Such themes are especially prominent in the descriptions given by Constance Lytton, who detailed her experiences with the tube in both the national and international press in January 1910, utilizing language that was implicitly critical of the medical procedures involved. She was forcibly fed twice a day with a stomach tube whilst imprisoned. Lytton explained that a large metal gag had been used to force open her mouth, an operation that had caused her to vomit repeatedly. She depicted her experience as:

> A living nightmare of pain, horror and revolting degradation. The sensation is that of being strangled and suffocated by the thrusting down of a large rubber tube which arouse great irritation in the throat and nausea in the stomach. The anguish and effort of retching into the stomach and the natural writhings of the body defy description.

Implicit in her narrative is the claim that a problematic form of medical technology was being employed upon her in a manner which had no discernible benefit for her health. Instead, she insisted that it was applied in order to subdue her and render her more compliant. She portrayed the experience as being so unnatural that her body revolted against it, yet the medical attendant persisted in the procedure.

Lytton went on to note that forcible feeding had definitely worsened her health, maintaining that on one occasion she had been repeatedly forced to digest food until she vomited so hard that 'the result seemed to surprise and slightly alarm the doctor and he called in an assistant to test my heart'. Even at this point, she alleges that further negligence took place. The medical assistance given to her was described as insufficient, with only a very brief and superficial examination being undertaken. She compared this to a similar test that she had had in Newcastle, but which was for purely medical purposes, where a specialist had tested her heart with 'elaborate paraphernalia' for ten or fifteen minutes.[83]

The experiences of Laura Ainsworth were publicly described by Emmeline Pankhurst (1858–1928) and depicted similar themes. The prying open of the jaws with a cold, steel instrument shoved between her teeth, followed by the insertion of a tube whilst Ainsworth was firmly held down is prominent in her portrayal of Ainsworth's situation. Nowhere in the accounts can be found even the slightest hint of medical professionalism or a desire to induce anything but cruelty. The physical encounter between the body and the medical technology was portrayed as bereft of sympathy and medical value. Furthermore, the compassionate traditions of the patient–doctor relationship appeared to have been abandoned. Negative health implications are also central to Ainsworth's account. The passage of the tube down her throat into her stomach was reported to have caused a choking sensation, leaving a horrible feeling of sickness. Pankhurst reported that the continued use of the tube rendered Ainsworth so weak that she became no longer healthy enough to submit to it, with the doctors then only being able to use a standard feeding cup on her.[84] Her entire body, it was claimed, had been reduced to such a weakened state so as to render her malleable to the desires of the prison medical officer, and to encourage compliance with the demands of the prison authorities.

The potential effect on mental health caused by forcible feeding was also raised later on in the militant campaign. Kitty Marion claimed that the physical and mental agony of her 232 feedings was so great that she had felt as if she wanted to put an end to her life by hanging herself on many occasions.[85] Meanwhile, the suffrage prisoner and trade activist William Ball was removed to a pauper lunatic asylum after enduring five and a half weeks of forcible feeding at Pentonville Prison, London. It was suggested that prior to this Ball had been 'of unusually good health', having been the champion runner of the Midlands, never once requiring medical treatment throughout the twenty years preceding his imprisonment, an allegation which implied the duality of the intentions behind the activities of prison medical officers.[86] Further controversy emerged when Emily Davies insisted publicly that the experience of being forcibly fed had affected her mental well-being to such an extent that she had attempted to commit suicide whilst in prison by flinging herself off the high iron railings outside her cell.[87]

The concept of the struggling patient clearly contrasts with the preferred images forwarded in advice given to practitioners on the use of gastric technologies where patients were deeply anaesthetized or given various trial runs at the tube until they became accustomed to its internal sensations. If anything, the forceful application of the tube seemed to justify claims surrounding the falsehood promulgated by laboratory scientists when attempting to justify their professional activities. At worse, accounts of forcible feeding held implications of instrumental rape. Forcible feeding was administered in such a way as to make it a physical and spiritual violation.[88] Similar claims had long been made regarding the employment of the speculum when diagnosing cases of syphilis. The speculum examination was perceived as voyeuristic and degrading, and one that inflicted mental and physical pain on the female sufferer, who became transformed into an innocent victim of male lust and tyranny. The brutality of the doctors was also commented on in these procedures.[89] There seemed to exist an underlying implication that the suffragettes were being penetrated in a sexual manner with the stomach tube. Contemporaries noted an uneasy similarity between the devices made to hold women for sexual pleasure and the tables and chairs, replete with stirrups and straps, which rendered women ready for the experience of forcible feeding.[90] Was it possible then, if the speculum examination implied vaginal rape that the stomach tube was constructed as the oral equivalent? Such themes come across in Lilian Lenton's description, within which she claims that the doctors:

> Amused themselves trying first one tube, then another, over and over again, pushing tubes, obviously far too thick, as far down as they could make them go, then, with excellent logic, trying thicker ones. This amusement they varied by pushing down small ones which I promptly coughed up into my mouth. Just as one tube reached my throat one of the doctors pushed his fingers down to keep it from coming into my mouth.[91]

The medical profession were not united in their support of forcible feeding and the militant suffragists were able to obtain the opinion of a wide range of medical authorities to justify their claims. One article published in *Votes for Women*, the leading organ of the militant Women's Social and Political Union, claimed that laceration of the throat was highly likely should the patient struggle. Furthermore, the hard parts of the tube might come into contact with parts of the stomach resulting in serious injury. There was also a risk that during either the insertion or removal of the tube into the throat, some of the food might enter the lung, leading to pneumonia. It was even claimed that there had been one case in an asylum where the patient had partly bitten off his own tongue after it had become twisted behind the feeding tube.[92]

Many asylum attendants seem to have viewed allegations of brutality and torture with some perplexity, and were more inclined to refute them. One anonymous author wrote to *The Times* in 1909 claiming that he had regularly employed the device for around forty years, both in hospital and private practice, and had never known resistance to be offered by the patient, excluding severe cases of insanity. He also stated that the gag that formed a crucial part of the procedure and which seemed to be causing particular concern, was in fact a simple contraption often used by dentists to keep the mouth open. He described outcries about the procedure of forcible feeding as 'indefensible and absurd'.[93] Dr George Robertson, physician-superintendent at the Royal Edinburgh Asylum, also argued that it had 'been a source of perplexity and astonishment' to those engaged in caring for mentally ill patients to learn that such methods of artificial feeding were being mistakenly represented as dangerous forms of torture, claiming to have performed the operation over 2,000 times without encountering any problems.[94] Similarly, in 1909, an anonymous contributor to the *British Medical Journal* argued that the instruments were safely used in prisons and asylums on a daily basis, and were even regularly employed by the patients themselves via auto-lavage.[95]

However, it was those engaged with physical, rather than psychological treatment who spoke out most strongly as it was the authority of their discipline which was the most in question. They also protested strongly that when employed on a *struggling* patient, the risks of forcible feeding were heightened in comparison to their use in an asylum. The esteemed London surgeon Charles Mansell-Moullin (1851–1914) wrote to *The Times* soon after the policy of forcible feeding was instigated, arguing that the 'hospital treatment' offered to the suffragettes constituted violence and brutality.[96] Shortly after, Dr Forbes Ross (1867–1913), surgeon at Kensington Hospital, London, also suggested that the methods of forcible feeding used in Birmingham Prison were an act of brutality beyond human endurance, going so far as to suggest that its use was frequently followed by the development of chronic pigmentary colitis, an intractable disease.[97]

Vivisectionists, too, became alarmed by the procedures, perhaps concerned with the implications regarding their ethics embedded within the claims of the suffragettes. The British surgeon and arch-vivisectionist Victor Horsley (1857–1916) condemned requests made by politicians for medical professionals to be asked to comply with problematic medical procedures that risked casting disrepute upon modern medical procedures. Speaking on the subject of one suffragist prison death, he claimed:

> So little did the Home Office do towards 'saving her life' as suggested in the Judge's question, that she became weaker and weaker until, finally, to avoid a terrible scandal they, to really save her life, ceased her so-called 'medical' treatment and turned her ill and suffering, out into the streets of Birmingham without even the means to get to her house.[98]

These representations of the stomach tube as the symbolic instrument of medical torture therefore constituted a climax in debates surrounding the clinical benefits of scientific medicine. It also brought forth questions that were central to the problematic discipline of laboratory physiology. The use of gastric technologies provided the most publicly renowned example of the direction of modern medicine, representing a culmination in fears regarding the intentions of members of the profession in pursuing such forms of behaviour, and the most negative aspects of the intrusion of laboratory medicine into the clinical level.[99]

This controversy did not go unnoticed amongst the public. Tellingly, one patient wrote to the *Manchester Guardian* at the height of the controversies in 1909. He seemed particularly unconvinced by the claims of the medical community of the apparent ease in which the tube entered the body. He detailed how he had been sent to Manchester Royal Infirmary some years earlier for various stomach tests that involved the use of the tube and the pump. He wrote that:

> My experience was such that the mere recollection brings back the sense of suffocation and nausea that accompanied the endeavour to pass the tube into the gullet. In spite of the patience of doctor and nurse and my own utmost co-operation it was fully a minute before I could get the horrible slimy thing down my throat, whilst the sweat poured down my throat. Yet both doctor and nurse were surprised to learn that it was the first time I had undergone the treatment, since I 'took it so well' and had not actually been sick.

The anonymous author stated that the revulsion which he felt was not the result of an exceptionally sensitive nervous system, and that he was in good health overall. He also expressed concern that for a patient with a condition such as a weak heart having to endure the instrument being forced into her throat violently must be 'dangerous in the extreme'.[100]

Conclusions

The incorporation of new scientific methods to treat gastric illness in the clinical setting was clearly problematic. On one hand, physiology had been successful in determining the nature of the digestive processes. Yet the full application of its associated techniques in the clinical setting was thwarted by the question of the necessity of induced pain inherent in both the experiments that preceded medical findings and in the clinical scenario. Antivivisectionists and militant suffragists utilized gastric technologies as a prime example of the new strand of cruelty that had permeated late nineteenth-century medicine. This was especially effective given the profound anxiety regarding gastric health in Britain. Sham feeding and forcible feedings were amongst the most prominent emblems of these. When set against a backdrop of perceptions of a nation beset with gastric problems, the utilization of gastric technologies further pointed to the alleged inefficacy of new procedures, and the inability of laboratory medicine to provide a cure for a problem that, for many observers, threatened to cripple the British nation.

4 THE SURGICAL STOMACH: BERKELEY MOYNIHAN'S FORGOTTEN SURGICAL REVOLUTION AND DUODENAL ULCER DISEASE, *C.* 1880–1920

While the future British King George VI (1895–1952) was serving in the Royal Navy during the First World War, he became plagued with gastric problems. Being at sea seemed to have intensified his long-standing abdominal complaints. This was potentially problematic given that the presence of a prince in the navy formed a integral element of the monarchy's traditional relationship to war. Invaliding him out was therefore not a decision to be taken lightly. Accordingly, his appendix was removed in August 1914 to cure him of his afflictions, and also with the hope of allowing him to return quickly to military service.[1] However, he continued to be crippled by agonizing gastric pains following the procedure and was eventually diagnosed as suffering from a duodenal ulcer in August 1916. A lengthy period of rest alleviated many of his symptoms, yet he became increasingly weak once again upon returning to sea duties in May 1917. Eventually a surgical operation was performed to remove the anomaly on 29 November 1917.[2]

Yet twenty years earlier, Prince George would have been highly unlikely to have been found suffering from a duodenal ulcer. In fact, the complaint was then rarely diagnosed. Even as recently as 1903, the *Edinburgh Medical Journal* had claimed that while 'cases of duodenal ulcer occur from time to time, still the disease cannot be called common'.[3] However, by the time that the prince underwent his operation in 1917, it was generally recognized as an important, unique disease category with significantly high incidence levels in Britain. Was this because the disease did not exist to any notable extent prior to this period? Or was it instead because members of the medical profession began to ask new questions intended to generate a new set of skills and knowledge which enabled them to recognize its symptoms with more clarity than before?

Abdominal surgery clearly played a central part in the story of Prince George's illness. The future king underwent two operations to cure medical problems that had not been recognized to any notable extent by earlier generations of surgeons and physicians.[4] The therapeutic history of George's illness is therefore evident of a sharp break in the management and treatment of gastric disorders in that it necessitated an abdominal operation. Recent changes in the area of surgical diagnosis and therapeutics had proven to be dramatic.[5] If an engagement with laboratory medicine and the adoption of associated techniques of gastric analysis had been characterized by caution, wariness and even outright hostility in Britain, the intrusion into the abdomen by turn-of-the-century surgeons reveals a sharp contrast.[6] Armed with a striking assurance in their ability to observe, manipulate and cure problems deep inside the body, surgeons were now able confidently to assert the validity and safety of their field, and to portray it as constituting a revolution in therapeutics.[7] The discipline of abdominal surgery was therefore well positioned to add a new element to the ongoing discourses surrounding the stomach in existence between medical professionals.

Encouraged by shifting ambitions and technical capabilities, a visible increase in the confidence of surgeons to manage the stomach occurred throughout the closing decades of the nineteenth century.[8] Ultimately, this encouraged both surgeons and physicians to identify duodenal ulcer with greater frequency and for them to conceptualize it as a far more common disease than had formerly been recognized. In fact, the condition became held up as a flagship disease, epitomizing the seemingly revolutionary speed at which surgeons perceived their capabilities to be advancing. It also encapsulated their new-found abilities to replace what they argued to be highly inaccurate knowledge of gastric diseases accumulated by nineteenth-century pathologists. Furthermore, the dissemination of new information regarding ulcers of the stomach and duodenum helped to strengthen the validity of the method of 'pathology of the living', whereby *live* internal human material was rendered visible to the surgical operator in a manner which it had not been before.[9] This had a profound influence on the nature of professional relationships within medicine.

Nineteenth-Century Surgery and the Abdomen

Early nineteenth-century surgeons, most notably John Abernethy, had been successful in popularizing models of the body that posited the stomach at the centre of an interactive biological system, depicting the organ as the key component of a highly complex network of sympathetic interactions.[10] This had equipped surgeons with the resources with which to carve out a new role in the management of internal disorders, rather than the external alone. Yet the surgical works produced throughout this period had always prioritized prevention above cure, and rarely put forward techniques involving the direct surgical manipulation of the stomach.[11] Furthermore, although ulcer of the stomach had become increasingly presented as a manageable disease, the aims of laboratory medicine in providing diagnostic and therapeutic regimes for the complaint had not been fully realized. Hence, there existed scope for surgeons to express a strong desire to demonstrate their ability to manage the condition.

Intrusive abdominal surgery had previously been recommended only for war wounds or for emergency situations such as the removal of accidentally swallowed foreign objects where the patient seemed to be at a high risk of death if intervention was not urgently undertaken.[12] A telling example occurred in 1824 when a man named C. A. Dempster, following a show of juggling tricks presented to his friends, attempted to swallow a sharp nine-inch table knife. Although he did not seem to be in pain, the object remained lodged in a precarious location in his stomach and required immediate removal. His journey from Carlisle to London kept the national press entertained for weeks. His initial journey to Manchester involved a trip in a jolty carriage, causing the patient excruciating pain. He then embarked instead upon a canal boat which worsened his symptoms, meaning that he had to disembark at Middlewich, Cheshire, where he died. Upon dissection, the knife was found to be partly dissolved, and it was lamented that Dempster should have remained still rather than seeking surgical intervention, as the knife might have dissolved entirely. *The Times* observed that the case had 'excited considerable interest in the public mind, as also in that of the profession' and reported the story in great detail, not least because the famed surgeon Sir Astley Cooper (1768–1841) intended to operate on Dempster.[13] However, such cases are highly suggestive that abdominal surgery would only be utilized when chances of survival were already slim or where some observable threat to life was in place. Even then it would only be undertaken by the most experienced surgeons.

The only major abdominal operation introduced during the early nineteenth century was Cooper's tying of the abdominal aorta to treat aneurism, a procedure first successfully undertaken in 1817.[14] Even the development of anaesthesia and pain-free operations during the 1840s failed to stimulate a rapid intrusion into the most inner regions of the body.[15] Instead, mid-nineteenth-century abdominal surgery became characterized more by caution and restraint than by surgical enthusiasm. Tellingly, when the general practitioner and medical author Robert Druitt (1814–83) published *The Principles and Practice of Modern Surgery* in 1852, he limited his suggestions for surgical intervention to cases of paracentesis abdominis, gastrotomy for obstructed bowels; and ovariotomy for ovarian tumours.[16]

Charles-Emmanuel Sédillot (1804–83) had proselytized gastrostomy in the 1850s, an operation involving the surgical opening of the stomach to bring relief to patients experiencing difficulty swallowing which was first undertaken in Britain in the same decade by John Cooper Forster (1824–96).[17] Yet the majority of abdominal procedures developed over the following decades focused upon diseases of women. Lawson Tait (1845–99), professor of gynaecology at Queen's College, Birmingham, went so far as to claim that the evolution of abdominal surgery had proceeded entirely from this field of enquiry.[18] Ovariotomy, the surgical incision into the ovary, became an established procedure of the abdominal area despite being surrounded by controversies over its validity.[19] A number of operations had been developed by the 1880s, such as hysterectomy, whereby the uterus was completely removed, and became a fairly common procedure in Britain.[20] Operative interference in the abdominal region focused initially upon reducing the influence of the reproductive organs on the general health of the female, as it was these organs which were believed to lead to a wide range of problems throughout the nervous system.[21] However, many of these procedures had fallen into disrepute by the end of the nineteenth century on the basis that they were deemed to be essentially mutilating rather than curative.[22]

Various factors encouraged operative restraint on other regions of the abdomen. Pre-Listerian surgeons were wary of the infections that might result from the exposure of the area.[23] The potential dangers involved in the introduction of new surgical techniques had always had to be weighed against the severity of the medical complaint itself and its immediate danger to life, not least because, like any untested medical innovation, there existed an unknown degree of risk to human life. Around mid-century, it was generally accepted that if a patient's constitution had become so weakened by factors such as intemperance and dietary neglect then the body's natural powers of reparation might have become entirely destroyed. This rendered such people as unsuitable candidates for highly intrusive abdominal operations.[24]

Fear of traumatic fever also made many surgeons hesitant to intrude too far into the abdomen.[25] Furthermore, wariness regarding the safety of anaesthesia, particularly in operations requiring the opening of the cavity for lengthy periods, seems likely to have played a significant part.[26] Surgeons noted early on that chloroform disturbed the stomach even when used for operations far away from the organ, causing continual nausea and vomiting which might ultimately prove fatal should it over-exhaust the patient.[27] Accordingly, it tended to be used principally for abdominal wounds rather than to provide cure for illness.[28] Yet, although fears over the potentially fatal effects of chloroform persisted throughout the century, more detailed studies published in the 1890s concluded that it was a safe substance if not overused. Confidence in chloroform increased, not least because it required the application of relatively small amounts in comparison to ether.[29]

Yet the main barrier to a speedier adoption of stomach surgery lay in the difficulties inherent in clearly distinguishing between the differential symptoms of abdominal diseases. Back in 1852, Robert Druitt had complained of this, exhibiting wariness in the pages of his surgical writing devoted to the abdomen, stating that

> our author does not wish his readers rashly to lay open the abdominal cavities of their patients, more especially when it is considered how uncertain are our means of diagnosis as to the nature and seat of the obstruction. He merely wishes to hint at a possible and desperate means of relief that might be adopted in some desperate case.[30]

Discussion of the relationship between particular gastric symptoms and specific diseases might rage for decades, as exemplified by the example of the complaint that was to be termed appendicitis. A sixty-year debate ensued about whether or not the identified symptoms were intra- or extra-caecal, but was eventually determined as being neither of these, but as being located in the ileo-caecal appendix. This led to the loss of the term 'typhilitis' which technically referred to inflammation of the caecum, and this was replaced by 'appendicitis'.[31] Overall, surgical interference in the abdomen occurred at only a gradual pace throughout the third quarter of the nineteenth century and a lack of diagnostic skill when dealing with the gastric system was an important reason for this. This meant that surgery then played a negligible role in the formulation of new methods of observing and treating gastric illness.

The Surgical Revolution

In sharp contrast, surgical operations within the gastric tract developed rapidly in Britain from around the 1880s, constituting what was soon claimed to be a revolution in the management of abdominal conditions. A variety of operations for the abdominal region were developed from this period onwards intended for men and women alike. This meant that problems which once fell under the realm of clinical medicine became redefined as surgical, a scenario with ramifications for professional activity. Nephrectomy, for instance, a procedure involving the removal of a kidney, became increasingly viewed as a safe and effective form of intervention.[32] The introduction of kidney surgery was also notable as it demarcated a moment when surgeons began to write texts aimed at practitioners which not only intended to disseminate information relating to the problems of the organ, but which also identified areas where the physician should no longer attempt to intervene. For instance, David Newman of the Western Infirmary, Glasgow, published his *Lectures to Practitioners on the Diseases of the Kidney Amenable to Surgical Treatment* in 1888, within which he recommended that movable kidney, renal affections and tumours of the kidney all required operative rather than medical intervention.[33] Hence, certain varieties of abdominal illness became reclassified from medically managed problems to surgically managed ones, forging new definitions of what constituted a 'surgical disease'. This ultimately impacted upon the interaction between medical disciplines.

Increasingly ambitious abdominal operations on ulcers of the stomach became a routine part of surgical practice from the 1880s onwards, although most of the techniques involved had stemmed from European countries. In Germany, Theodor Billroth (1829–94) was a crucial figure for the professional advancement of abdominal surgery, despite the sacrifice of many lives before his practices and post-operative care became fully refined.[34] Billroth was directly responsible for the introduction of a number of new operations including the first oesophagectomy in 1871 and the first laryngectomy in 1873.[35] What became known as the Billroth I operation involved the resection of the distal human stomach and anastomosis to the duodenum. The Billroth II procedure, meanwhile, involved closing the top of the duodenum and connecting the resected stomach to the jejunum. However, he became most famous for performing the first successful gastrectomy for gastric cancer in 1881, although again only after many ill-fated attempts. This operation involved the removal of a portion of the stomach fourteen centimetres in length.[36] This represented the first major intrusion into the organ by nineteenth-century surgeons, and its inclusion in surgical literature constituted a radical departure from the theoretical suppositions forwarded by early nineteenth-century surgeons such as John Abernethy.

Opening the abdominal cavity to treat ulcer of the stomach was first proposed and undertaken in Britain by the Bristol surgeon Nelson C. Dobson in 1883, who began to treat perforated ulcers simply by sewing the lesion shut or, alternatively, by stitching it to the abdominal wall. Both of these procedures were thought to encourage natural healing.[37] These operations became increasingly accepted as an appropriate form of treatment. A mounting number of successful case studies began to appear in the medical press from the early 1890s onwards.[38] From thereon, confident surgeons argued for the general adoption of such methods, condemning what they perceived to be wariness and caution amongst many of their colleagues. In 1892, Dublin surgeon Alfred R. Parsons (1865–1952) declared that it was an embarrassment that abdominal perforation should still be a cause of death in an organ that had by now been frequently incised, partly excised and even completely removed in some animals, going on to ask, 'in this, the age of gastroectomies, gastrostomies, gastroenterostomies, pylorectomies, are these specimens not a reproach to our diagnostic skill, or operative courage?'[39] He went on to argue that there was great potential for the development of new techniques, asserting that now was the time to treat ulcers of the stomach *before* they reached potentially fatal stages.[40]

The issue was raised infrequently throughout the following years. The Guy's Hospital surgeon Hastings Gilford (1861–1941) stated in 1898 that less than ten years earlier, perforation of an ulcer of the stomach would be treated medically with little hope of success. Yet the results of contemporary surgical treatment seemed to him to be so encouraging that he anticipated that in the future up to 50 per cent of lives could be saved as a result of surgical intervention. He also asked whether it was the time to consider operation before the ulcer progressed into potentially dangerous stages such as haemorrhage or perforation, assigning any previous deaths to the effects of peritonitis rather than a direct consequence of the opening of the abdomen.[41]

Gilford's question pointedly raised issues regarding the extent to which surgery should intrude into the inner body, and whether caution and restraint were in fact justified. It is also indicative of surgery becoming far more certain about its therapeutic abilities. Assertions were made that by the first decade of the twentieth century abdominal surgery had *revolutionized* internal therapeutics, a process no doubt stimulated by the increasingly high social esteem that the heroic surgeon came to hold around the turn of the century. By then, the ability to treat diseases and conditions that had for centuries remained incurable might elevate the surgeon to a role of national hero. This was exemplified by images of the throngs of crowds who went to Waterloo Station to dispatch the prominent British abdominal surgeon Frederick Treves (1853–1923) to the Boer War.[42] Yet, it is notable that in fact only one new operation was introduced between 1900 and 1906. Progress seemed to lie instead in improvements in methods and techniques and a diminishing death rate.[43]

The most vocal character at this time espousing notions of a surgical revolution in the abdomen was the highly influential Sir Berkeley Moynihan (1865–1936) who became the leading advocate of British abdominal surgery. Moynihan began his career as house surgeon at Leeds General Infirmary and went on to become a world-famous operator. His reputation spread internationally, being significantly enhanced following his invitation by the American Surgical Association, the senior surgical body in the United States of America, to read a paper at its meeting in May 1903. This experience enabled him to form lasting friendships with surgeons who were at the helm of the discipline, including the esteemed brothers William Mayo (1861–1939) and Charles Mayo (1865–1939), as well as George Crile (1864–1943). It also led to joint publications on abdominal surgery that became internationally renowned. His prominence in Britain became increasingly raised following his election to the council of the Royal College of Surgeons in 1912 and his six-year presidency that commenced in 1926. Moynihan was knighted in 1912, appointed CB in 1917 and KCMG in 1918. His other honours were numerous, the most notable being that of baronet in 1922. He was also awarded several honorary doctorates and society memberships.[44]

More so than any of his colleagues, Moynihan was keen to present the work of abdominal surgery as constituting a revolutionary break in medical diagnosis and therapeutics, stressing that breakthroughs were occurring at an extraordinarily rapid pace. For instance, in the second edition of *Diseases of the Stomach and their Surgical Treatment*, published in 1904, Moynihan and the esteemed Leeds surgeon William Mayo-Robson (1853–1933) stated that in the previous edition, printed just two years earlier, they had 'ventured to forecast that in the near future there would be a great activity in the surgical treatment of the diseases of the stomach' and that 'this forecast has been abundantly justified'. The number of papers, discussions and addresses dealing specifically with the surgical diseases of the stomach was presented as being so remarkable that it proved difficult even for the authors to keep up with the subject's literature.[45]

Moynihan went so far as to depict the rapid advance of contemporary abdominal surgery as the next stage of a revolutionary process that had commenced with the discovery of germs and antiseptics decades earlier.[46] For him, invasive surgery's potential to observe and treat complaints hidden deep inside the patient heralded the commencement of a final stage of a long medical revolution. Moynihan's viewpoints on the contributions of his discipline were exemplified in a work written late in his life, entitled *The Advance of Medicine* (1932), in which he provides a narrative of medicine from ancient to modern times, but with a hefty, and audacious, emphasis upon turn-of-the-century abdominal surgery, a topic which constituted approximately half of his book. Within this text, he concluded by stating that further medical progress was in fact not possible, as the recent accomplishments of surgery were such that they could never be improved upon.[47] What this represented was an eagerness to overturn what were presented as the failings of earlier forms of gastric therapeutics in fully managing digestive problems. Management of ulcer of the stomach and duodenum therefore became central to a far wider debate, coming to act as an integral part of a wider stratagem of constructing a surgical revolution. This added a compelling new element to the complex interaction in existence between medical disciplines, and led to the prioritization the management of the stomach as an essential facet of their agenda of professionalization.

Abdominal Surgery and Scientific Research

Despite such audacious claims of success, operative surgery still faced problems of acceptance by physicians, general practitioners and patients, all of whom maintained that there existed dangerous post-operative problems, the stigma of scars, delayed consequences and relapses. The formulation of such criticism was part of a professional response to the emergence of a confident discipline which was busy redefining traditional areas of the work of the physician into surgical areas. One way by which surgeons strove to validate their new approach was to present it as one with an intrinsic ability to produce important new strands of patho-physiological research.[48] Abdominal surgeons proved to be far more successful than physiologists in providing a convincing rationale for this role. The ability of contemporary surgery actively to observe and visualize living internal phenomena, rather than only morbid, was the key to this. Surgeons presented themselves as being in a particularly suitable position to obtain fresh insights and observations on medical complaints. This imbued the discipline with the potential to make a vast array of corrections to pre-existing medical knowledge.[49] A further aspect of this process was to maintain that all other methods of internal exploration so far developed were much less precise. Morbid anatomy became the prime target of this, not least because in many ways it had made the strongest contribution to scientific knowledge of the gastric system during the preceding century.

Close associations were, however, initially forged with laboratory medicine. This was in many ways natural, given that both fields shared a similar approach to the human body and its medical conditions, both engaging with inherently localized and reductionist viewpoints of the patient's body.[50] Furthermore, both disciplines were highly ambitious in convincing the public of the scientific validity of their work. Yet, if anything, the therapeutic gains of surgery were far more plentiful than those of physiology, providing scope for the undertaking of operations that might easily have been dismissed as reckless and dangerous were it not for their alleged scientific foundations. Certainly, surgery did not evoke the same sense of disdain that laboratory medicine did throughout Britain, not least because its therapeutic rationale was more convincing. Hence, although surgeons often closely followed the discoveries made in the laboratory, the struggle to prove the validity of new medical procedures was not tainted by the same connotations of medical brutality and cruelty undertaken on animals. It could be more easily justified by claims of cure. Furthermore, by this period, controversies relating to surgical mutilation had become less prominent than they had been in the mid-nineteenth century. Yet developing their own mode of scientific investigation helped to distance the field further from association with vivisectionism and medical brutality

An approach shared by the two disciplines was the dislodgement of culturally pervasive ideas which emphasized the stomach's essential internal significance. They attempted to achieve this by paying closer attention to the various regions of the organ, as opposed to emphasizing the bodily functions of the stomach as a whole. The laboratory scientist, with his inherent interest in the more minute areas of the body, persistently maintained that the role of the organ had for too long been assigned highly exaggerated physiological importance. In certain strands of medical thinking, the organ's functions effectively became reduced to a similar role as the hump of the camel, acting as a larder where digested food was kept aseptic and sweet prior to it passing through to more important digestive stages.[51]

The impact of this reductionist approach upon surgical thinking about the stomach became clearly discernible. Both surgeons and physiologists began to claim that man could exist healthily without a stomach. Surgeons advocated new operative possibilities involving the removal of large parts of the stomach, or even, in some cases, the entire organ. These procedures were supported by medico-scientific research which claimed to have proven the relative insignificance of the stomach. The most influential experiments undertaken on the organ were those of the Russian physiologist Ivan Pavlov (1849–1936), a chief advocate of ideas that the organ could be completely removed with no negative health effects. This idea was derived from one of his experimental dogs having lived in good health for almost five and a half years, completely stomachless.[52] In contrast, experiments made in Naples towards the end of the 1880s involving the extirpation of the pancreas from laboratory dogs and cats had suggested that it was the smaller abdominal organs that were more essential to general health as these animals had suffered from poor health and died. For instance, diabetes was observable when the pancreas had been surgically removed.[53] Yet no specific problems or risk to life seemed to have arisen in the stomachless laboratory animal or human patient.

The possibility of completely removing the human stomach became increasingly perceived as feasible.[54] In 1897, the *Lancet* described a successful operation performed in Switzerland involving the complete amputation of the stomach of a fifty-six-year-old woman suffering from diffuse cancer. Her operation involved the stitching of part of her small intestine to her oesophagus in order to maintain the continuity of the digestive tract. She was reported to have recovered so well that she was presented to an audience at the meeting of the Central Association of Swiss Medical Practitioners the following year. In 1898, the *Lancet* also reported a case of a fifty-two-year-old German woman whose entire stomach had been surgically removed. She had suffered from excruciating stomach pain and weight loss for over a year. The procedure lasted only seventy-five minutes and it was reported that she was able to drink a glass of claret that evening. Only a fortnight later, she was able to digest potato soup and grated ham, and was fully mobile after fourteen days.[55] Such conceptualizations of the stomach are indicative of attempts to portray the organ as obsolete; an evolutionary remnant no longer necessary for human existence. Take away the stomach, or parts of it, and it seemed that its problems would simply disappear with minimum consequences for a patient's health and longevity. Of course this presented a radically different perspective on the stomach and its medical conditions to that typically presented by dieticians and physicians. And this brought forth drastically different procedures to earlier traditions which had focused intensely on prevention and adjustment to external environments.[56]

From the first decade of the twentieth century, surgeons not only chose to draw upon existing physiological principles, but also began to claim that the ability of surgery to observe living human material at the operating table might in fact also enable the surgeon to develop his own unique strand of research and scientific observation. Abdominal surgeons began to assert that their work too held the potential to mirror, if not improve upon, the physiologist's quest for knowledge as well as cure.[57] Certainly, it seemed that the surgeon had access to something that the pathologist, who could observe the interior of the dead person only, and the physiologist, relying upon the removal and analysis of gastric contents without any immediate observation of a disease, had never had: the ability to open the living body painlessly, observe living human material and record a variety of observations about health and disease.[58]

Whilst operating upon the interiors of the living human body, the abdominal surgeon could observe a variety of conditions, being well-positioned to glean important knowledge of diseases and to elucidate unexpected connections between diseases. The operating theatre, it was hoped, might become just as much a place of learning as it was a place of cure.[59] Ideas were asserted that the surgeon when at work in the theatre should no longer be content with narrowly regarding himself as a technician but should instead consider himself as an experimental biologist carrying out valuable research into the living processes of human bodily disease.[60] Certainly, for Moynihan, the operating theatre appeared to be such a potentially procreant laboratory that he later declared that 'our knowledge of all abdominal diseases has been so greatly and so quickly changed by operative research that a complete revolution in thought and in action has resulted'.[61] Furthermore, by producing their own experimental models, surgeons were able to distance themselves from the problematic connotations associated with laboratory medicine that were proving to be so contentious amongst the British public. After all, their findings did not necessitate the infliction of pain on the animal. Although the human body itself was, in a manner, being experimented upon, this experimentation essentially formed a secondary part of their therapeutic intervention and could be justified by the diagnostic and therapeutic potential which might result.

The concept of the 'pathology of the living', a term coined by Moynihan to describe the observation of living phenomena in the theatre, was created with the intention of encouraging the surgeon not simply to remain content with providing a cure. As he later explained:

> When I coined the phrase 'the pathology of the living', I did so in order to show that surgery might be used not exclusively as a therapeutic resource, but also as a means of inquiry. And the results of this inquiry have already been stupendous and revolution-

> ary. This wide and detailed examination which attempts the exact correlation of the symptoms with the pathological conditions discovered, constitutes a form of research than which few can be more difficult or bewildering.[62]

Of course, it is important not to assume that this concept remained unchallenged or that it was introduced unproblematically. Like physiology, the rise of the field of abdominal surgery constituted a threat to traditional professional discourse surrounding the stomach. Certainly, Moynihan persistently felt the need to stress the value of his new frameworks of enquiry, being concerned that the surgeon's findings were more likely to be viewed as valueless as they tended to oppose completely 'the doctrines held as sacred by the pathologist'. He warned that the research findings of abdominal surgery were often being completely ignored or rejected by many members of the British medical community at the expense of trust in the generally accepted findings accrued from nineteenth-century pathological anatomy.[63] Accordingly, his criticism of earlier investigative methods was often fierce. On one occasion, he lambasted nineteenth-century pathological enquiry as 'shackles by which he [the surgeon] has been fettered for so many years', challenging those who disagreed with him to compare the accuracy of the findings of contemporary surgical research to nineteenth-century pathological results.[64]

By scientifically exploring the stomach and duodenum, confident surgeons discovered a particularly suitable object of enquiry whereby rich and dramatic corrections could be made to the existing body of knowledge of the inner regions. This was because knowledge of the organs, up until now, had remained relatively vague and therefore open to conjecture. The pathology of the living, it was suggested, had completely reconfigured knowledge of even the most basic pathological facts relating to the stomach, even on the most fundamental of matters including its shape and position. Tellingly, in *Diseases of the Stomach and their Surgical Treatment* (1904), Mayo-Robson and Moynihan convincingly insisted that the collapsed, flat-walled and flaccid bag regularly depicted as the empty stomach did not at all represent the organ's true condition, but instead illustrated a stomach examined after death that had been subject to post-mortem softening, relaxation and pressure. They went on to argue that the scientific methodology of the pathology of the living had enabled new conceptions of the stomach, and had proven it to be a highly extensile organ capable of considerable expansion and contraction that allowed scope for different amounts of consumed food. In conclusion, they suggested that 'the stomach is not an inert bag, hanging down when empty like a flapping sail, with its walls in contact, but an active, living organ, capable of expansion and contraction, which adapts the size of its cavity to the amount of its contents'.[65]

Even existing knowledge of the internal position of the stomach became challenged, accompanied by warnings that the presence of such a basic mistake had long led physicians to make a multitude of diagnostic and therapeutic errors. For Moynihan and Mayo-Robson, the standard description given in medical and anatomical textbooks depicted the cardiac two-thirds or three-quarters of the stomach as being placed vertically. However, they now claimed that this arrangement would in fact be physically impossible as, if it was actually achievable, it would cause the displacement of the pancreas as well as many other small internal organs.[66] Instead, they maintained that they had determined a more accurate description of where the stomach lay in relation to the other internal organs, a development made possible only by the ability of surgery to make observations of the living inner body. Early twentieth-century internal surgeons were therefore able to claim that the current phase of the envisioned long surgical revolution held tremendous investigative potential and that the rapid accumulation of this during the century's opening decade was illustrative of the aims of the current phase of the surgical revolution. Yet it also entailed an engagement with conceptualizations of the stomach that deprioritized its internal, sympathetic significance, presenting new ways of observing its medical complaints that formed a sharp contrast with those of other medical disciplines.

Surgical Diagnosis

A further important argument utilized to express the validity and necessity of abdominal surgery was claims of the accumulation of knowledge from surgical research which would prove to be diagnostically and therapeutically valuable in the clinical setting. Operations appeared to be useful environments where information gained of the early stages of disease not normally visible during the post-mortem examination could be accrued. These had always been hidden to the pathological anatomists who tended only to see the later, fatal consequences of medical conditions. It also enabled surgeons to claim superior research potential than the pathologist, who had, during the nineteenth century, tended to act only as a check upon clinical diagnosis, either correcting or confirming the diagnostician's conclusions. Although this role held potential importance, the pathologist had always remained relatively incidental to the clinical process, with his opinion being sought only after crucial decisions had already been made.[67] Turn-of-the-century abdominal surgeons were far from content with being employed in such a limited capacity, which is unsurprising given the strength of their ambitions. Accordingly, the surgical gaze turned increasingly towards the exploration of the earlier stages of disease that had remained invisible to the post-mortem examiner. By accruing such information, the abdominal surgeon hoped that he too might be able to contribute new factual evidence that could be correlated with the symptoms of the living patient, with the overall aim being to improve medical diagnosis generally, as well as to enhance the role of the surgeon within this. This also assisted them in avoiding the problem of accruing scientific knowledge without visible therapeutic benefit.

The diagnostic potential of abdominal surgery had occasionally been hinted at but had never been fully realized throughout the nineteenth century. This is not to say that diagnosis had not formed part of surgery prior to the period in question. If we return to the example of Abernethy, his work had certainly attempted to shift the focus of surgical attention towards the inner body in order to carve out new roles.[68] However, he had not been able to provide surgical possibilities for areas deep within the body. The turn-of-the-century abdominal surgeon, however, envisioned a much stronger role for members of his field. Tellingly, in 1896, The famed Glasgow surgeon Alfred Ernest Maylard (1854–1947) argued that

> it is not, I venture to think, too venturesome to predict that the day is not far distant when the stomach will be opened, explored and resutured *for purely diagnostic purposes* with as much freedom and security as is now done, for instance, in the case of the brain.[69]

Such viewpoints became increasingly prominent within medical literature. Mayo-Robson can be found suggesting in 1900 that cases of obscure abdominal pain producing chronic invalidism or debility definitely necessitated the opening of the abdomen in order to clarify diagnosis and to suggest appropriate therapeutics.[70]

The eagerness of the surgeon to open the inner body for diagnosis was often so strong that it was not uncommon for calls to be made for even non-life threatening cases of dyspepsia to be investigated with the knife.[71] However, objections were raised to seemingly extreme varieties of exploratory surgery, and requests for restraint became frequently asserted. As one critic suggested, 'opening the abdomen to see whether it ought to be opened' only to be confronted with a situation where 'there is little that is pathological to be seen in the stomach; but a few silvery adhesions here or there, or some enlarged glands in the omentum' was problematic.[72] Strikingly fierce opposition came from the Birmingham physician Robert Saundby (1849–1918) who argued in 1904 that no experienced physician who had dealt with diseases of the stomach would accept the opinion that cases of severe dyspepsia justified an exploratory incision. Instead, for him, surgery was best confined within certain imposed limits.[73]

However, surgeons were equally dissatisfied with the quality and accuracy of the clinical data provided by physicians prior to operation, not least because the precision of prior knowledge of the patient's illness might later prove to be a crucial factor towards operative success or failure. Success in itself was of course a factor necessary to prove the validity of invasive surgery to the general public as well as to the medical profession.[74] Yet many physicians remained wary of new operative procedures. Cautious practitioners, deterred by fears of a high risk of death, might only send their patients to the surgeon when their health had reached an especially low, life-threatening point. In response, surgeons began to produce accounts where cases of perforated ulcer were presented to them in a practically hopeless condition, with one telling example detailing 'the stomach wall being so rotten that it was actually torn to the extent of one and a half inches, and a large mass of tissue apparently malignant, occupied the anterior wall of the organ at some distance from the ulcer'.[75] Accordingly, increasingly vocal calls were made that the surgeon should not remain content to accept and act upon diagnoses given to him, but should go over the medical evidence himself, being prepared to supplement it with surgical insight and to make more accurate diagnoses where necessary.[76]

Part of this debate between medical disciplines resulted from the approach of the surgical diagnostician being remarkably different from that of the physician on a conceptual level. It forwarded approaches to the body which contrasted sharply with those favoured by the physician who, as we have seen, had long concerned himself with prevention via common sense dietary management. It also presented intrinsically localized models of gastric illness which neglected factors traditionally deemed essential by the physician, such as the relationship of the digestive complaint and the patient to socio-cultural environment or constitutional factors. The surgeon required much higher levels of precision if his operation was to be successful. This necessitated vastly different ways of conceiving the medical complaints of the organ, focusing not only on the disease itself but also upon the subtle differences presented by the same disease in different parts of the organ.

Accordingly, as in pathological anatomy and laboratory medicine, representations of the stomach presented within surgical texts partitioned the organ into highly specific regions. Enquiry shifted from the usage of terms such as dyspepsia towards the construction of a more clearly defined set of classifications that separated the organ into its component parts. Frederic Bowreman Jessett (1837–1927), surgeon at the Cancer Hospital, Brompton, argued that, from a surgical point of view, it became seen as highly important to differentiate between the symptoms of the organ in each particular area rather than between the abdominal organs, even when dealing with the same disease. For instance, he utilized the example of an operation being suitable for the relief of carcinoma affecting the pylorus as being completely inappropriate for the very same disease in the cardiac end of the organ.[77] Hence, Jessett did not structure his discussion of the stomach into analysis of its various diseases but instead divided into the more localized operative procedures of pylorectomy, gastro-enterostomy, combined pylorectomy and gastro-enterostomy, gastrostomy and jejunostomy.[78]

Alfred Maylard adopted a similar approach, noting that many surgical authors were failing to distinguish between the almost imperceptible differences existing between the symptoms of disease in one part of the alimentary tract and another, even when these occurred in distant parts. For him, the apparent ambiguity existing in the nomenclature of operations within the alimentary tract highlighted the necessity of revision, and the accumulation of a comprehensive consideration of the stomach. Rather than taking particular injuries or diseases, within his *A Treatise on the Surgery of the Alimentary Canal* (1896) he treated the subject regionally in the belief that this offered a better scheme for surgical treatment. Of course, this was not too dissimilar from the approach adopted by early nineteenth-century pathological anatomists. Yet surgery was able to offer even more precise analysis as it was undertaken on a living patient. This shift in diagnostic practice was, firstly, a product of shifting surgical ambitions and, secondly, a response to the wariness of other disciplines within the medical profession towards an emergent and ambitious form of surgery.

The Rise of Duodenal Ulcer

Abdominal surgery clearly transformed itself into a field which claimed to have a sound scientific background, yet one which also held its own research and diagnostic potential. Duodenal ulcer disease became the flagship disease of early twentieth-century surgery, epitomizing modern surgery's diagnostic capabilities and professional aspirations. This ensured that the complaint received particularly high levels of attention at the start of the twentieth century. Even at the end of the preceding century, many physicians and surgeons observed that duodenal ulcer was still being persistently misclassified as gastric ulcer. Few attempts had been made to distinguish between the pathology or potential differences in the symptoms of the two kinds of complaints. This could be problematic given that physicians might diagnose their patient with gastric ulcer or some other problem, and send them for an operation to cure this, only for the surgeon to find no lesion in the stomach. This led to surgeons raising warnings throughout the early 1890s that in order to improve the potential of saving the lives of those suffering from perforated ulcer more precise forms of knowledge than those accrued by nineteenth-century pathologists needed to be gained of the complaint.[79]

This was successfully achieved. By 1910, ideas relating to duodenal ulcer diagnosis seemed to have spread so rapidly that the *Dublin Journal of Medical Science* was able to ask:

> who of us ten years ago would have believed, had we been told it, that within such a short space of time duodenal ulcer would have emerged from the category of rare diseases and taken its place as one of frequent, nay of common, occurrence?[80]

Much of this attention to detail stemmed from the precision which abdominal surgeons now demanded. Even up until the late nineteenth century, surgeons persistently complained of an inability to distinguish diseases of the duodenum from other gastric illnesses.[81] Certainly, nineteenth-century writing on the duodenum had tended to be appended to literature on the stomach and was noticeably brief.[82] Tellingly, when the physician Samuel Fenwick (1821–1902) published *The Morbid States of the Stomach and Duodenum and their Relation to the Diseases of other Organs* in 1868, despite the promising title the text devoted just three pages to problems of the duodenum.[83]

The possibilities of providing more precise localized descriptions of diseases such as duodenal ulcer stemmed principally from the research agenda of abdominal surgeons and their emphasis upon partitioning the abdomen into concise areas. Berkeley Moynihan's *Duodenal Ulcer*, published in 1910, was the end result of over a decade of research and publications on the abdomen, and acted as an integral part of wider campaigns to convince the public and medical profession of the validity of invasive abdominal surgery. Moynihan suggested that of all the recent advances in the understanding of abdominal disease resulting from operative observation, there seemed to be none of greater importance than that of ulcer of the duodenum.[84] Hence, the elucidation of knowledge of duodenal ulcer served fundamental professional purposes for the turn-of-the-century surgeon as it strengthened contested arguments surrounding the role of modern surgery within medicine.

One aspect of Moynihan's reorganization of knowledge of the stomach and duodenum involved the assertion of claims that duodenal ulcer disease was in fact strikingly more common than assumed. This was presented as a surprising discovery made clear only as a result of recent surgical research, and as illustrative of the profound capabilities of abdominal surgery to discern important information regarding internal disease. In fact, the apparent superiority of surgical observation and examination seemed to have suggested that in earlier times the medical community was burdened with inaccurate views relating to duodenal ulcer disease's frequency and a subsequent misclassification of it as a gastric complaint. Moynihan maintained that duodenal ulcer had always been prevalent but had remained hidden from the medical gaze until surgeons opened up and investigated the abdomens of living humans. It was abdominal surgeons who had raised awareness of it as a condition of high significance, and in doing so were able to emphasize further the value of their work in correcting mistakes made earlier by alternative regimes of bodily enquiry.

This approach had become common with many abdominal diseases. Pancreatitis, for instance, was shown to be surprisingly prevalent, with the medical profession's attention again only being drawn to it as a result of the opening of the abdominal cavity.[85] Similarly, it was claimed that congenital hypertrophic stenosis of the pylorus had been shown to have high incidence rates once observations of it had become more frequently made on the operating table. For Moynihan, this condition had gone from being understood as 'one of the freaks of medicine having little more than pathological interest' to being 'a well-established clinical entity'. It seemed to him that in earlier periods, death from the disease had been persistently certified as death from marasmus, vomiting or intestinal obstruction.[86] Finally, 'hour glass stomach' had been discerned, a condition whereby the cavity of the organ seemed to be divided into two unequal sections, also referred to as 'double stomach' or 'bilocular stomach'.[87] Yet these were relatively obscure problems in comparison to duodenal ulcer disease. The complaint proved to be particularly useful because gastric ulcer disease was well known amongst both the medical profession and the public. Claiming duodenal ulcer to be a more prevalent disease than generally thought was not a particularly important enough point in itself. Hence, the complaint became presented as dangerous and life-threatening, with its commonality posing a prominent risk to the health of the community. In doing so, the validity of speedy surgical intervention could be more firmly justified.[88]

Once the abdominal surgeon began to announce surprisingly high frequencies of duodenal ulcer disease, then he found himself in a more suitable position to recommend changes in clinical diagnosis and therapy, thereby encouraging a wider acknowledgement of the importance of these newly elucidated disease entities throughout the general medical community.[89] Duodenal ulcer had been referred to only infrequently in nineteenth-century medical literature. Yet for Moynihan and his colleagues, this seemed attributable to the complaint's notorious difficulty to diagnose rather than a rising trend in its incidence.[90] Certainly, turn-of-the-century surgeons increasingly adopted the view that its symptoms had been so confused that it had been frequently misdiagnosed as dyspepsia or gastritis.[91] Moynihan suggested that in fact a host of nineteenth-century authors had taught information on gastric and duodenal ulcers which modern surgery was now revealing to be dramatically inaccurate.[92]

It became frequently alleged that morbid pathology had virtually ignored the duodenum. Tellingly, one author argued that only sixty-seven cases of duodenum diverticula had been recorded by 1910, but that this did not represent nineteenth-century frequency because, 'as a matter of fact, the duodenum is rarely carefully examined at the routine autopsy, and still less frequently has it been the subject of surgical intervention'.[93] Mayo-Robson also claimed that anatomical knowledge had mistakenly classified the duodenum simply as a division of the intestine, an error still being perpetuated at the start of the twentieth century. Furthermore, it seemed to him that ulcer of the duodenum had been wrongly considered as synonymous with ulcer of the stomach when in fact the two anomalies could now be proven to have strikingly distinct features.[94] Yet the living pathology of the duodenum was considered to have now been studied to such a level that more precise forms of diagnosis were possible, meaning that, as Moynihan claimed, ulcer need no longer masquerade under terms such as 'hyperchlorhydria', 'acid gastritis' and 'acid dyspepsia'.[95]

Moynihan adopted a particularly confident approach to diagnosis of the complaint, stating that 'there are few diseases whose symptoms appear in such a definite and well ordered sequence as is observed in duodenal ulcer'.[96] Hence, it constituted the prime example of the potential of surgical research to be transferred effortlessly into the clinical setting. Its early history was noted to involve a sense of weight or distension in the epigastrium after eating, a problem which occurred approximately two hours following a meal. Eating food whilst suffering from this pain seemed to bring some relief. This 'hunger pain' seemed to emerge at remarkably characteristic times. Moynihan also went on to state that chronic duodenal ulcer was characterized by the periodicity of symptoms and their recurrence from time to time in 'attacks' which often seemed to have been brought about by exposure to cold, wet feet, worry and overwork or a hasty, indigestible meal. It seemed more likely that the practitioner would observe duodenal ulcer patients in the cold winter months, whilst in summer these symptoms would be virtually absent. Furthermore, there seemed to be high levels of acidity in duodenal ulcer cases. Moynihan used this observation to assert his views that earlier diagnoses of 'hyperchlorydria' or 'acid gastritis' were most probably duodenal ulcers.[97]

The new diagnostic possibilities offered by abdominal surgery also challenged pre-existing assumptions regarding gender and gastric ulcer disease. Throughout the previous century, the disease had been construed as one which mostly afflicted young women, often anaemic maid-servants with symptoms of frequent vomiting and haemorrhage. However, duodenal ulcer appeared to be predominantly male, meaning that claims of its femininity became increasingly questioned.[98] Moynihan aimed to uproot the foundations of long-standing beliefs such as these by claiming instead that none of these symptoms were necessarily suggestive of gastric ulcer, due to them having been constructed without the examination of the early, inaugural symptoms observable only during the examination of the pathology of the living. Such assumptions appeared to rest upon details of the post-mortem examination of patients following the fatal perforation of an ulcer.[99] Instead, Moynihan identified the typical patient as being aged between twenty-five and forty-five and highly likely to be male.[100]

Even the generally accepted nature of duodenal ulcer disease as functional was challenged by Moynihan, who suggested instead that

> it is a very curious feature in connexion with this disease that perfectly accurate accounts of its symptomatology are given by authors who do not seem to have the remotest conception that the condition they are describing is not, as they suppose, one of 'functional' disorder, but one in which a demonstrable organic lesion is present.[101]

By suggesting that generations of textbooks had described the main symptoms of stomach problems incorrectly as functional he was able to present one of his strongest cases yet for the validity of surgical judgement. This enabled him to assert that although the vague, but commonly employed term dyspepsia had long been utilized to describe a large group of functional conditions, it was now revealed to be a variety of precise problems associated with organic digestive disorders and diseases. Given the apparent significance and prevalence of such complaints in Britain, the importance of the redefinition of a group of diseases with such high social and cultural standing should not be underestimated. Surgical research meant that organic stomach diseases could now be recognized at much earlier stages and given more precise interpretations, with the end result being, for Moynihan, that

> time will show that possibly all, certainly nearly all, of the cases of protracted and recurring 'dyspepsia' are due not to vices of secretion, though indeed these may be present, but to organic changes in one or other of the viscera.[102]

Professional Contestation

Peptic and duodenal ulcer disease were clearly allocated a prioritized role in the language and activities of surgeons around the turn of the twentieth century. It is therefore unsurprising that their treatment became an intense topic of debate and discussion amongst various sectors of the medical profession from this period internationally. Yet if duodenal ulcer was transformed into the flagship disease of abdominal surgeons, then it worked in the favour of physicians inclined to maintain their traditional authority in gastric therapeutics to challenge surgical views on the problem vigorously. In fact, gastric ulcer became a highly contested site as disagreement over its treatment escalated. Tellingly, in 1923, I. W. Wheeler, president of the Royal College of Surgeons in Ireland, described gastric ulcer as a 'no man's land' where 'attacks and counter-attacks to gain possession of this territory have been made' by both physicians and surgeons, a situation at odds with the best interests of the patient.[103] What ultimately emerged was a convolution of approaches towards treating the stomach, and little, if any, standardization amongst the medical disciplines.

Despite the lofty aspirations of abdominal surgeons, medical traditions of treating stomach disorders persisted. Physicians continued to insist, for instance, that ulcers could heal spontaneously without operation if the patient was given rest, diet and drugs. Surgeons, however, maintained that cases of permanent 'spontaneous healing' were rare, and that operation provided a far more complete cure. Advocates of laboratory medicine also complained that physicians still refrained from fully incorporating the findings of physiological knowledge into their treatment of stomach complaints, with relief being purely empirical or a 'shot in the dark'.[104] Conversely, physicians remained wary of new procedures and technologies. Radiographists, meanwhile, also turned towards attempting to manage stomach problems with their new techniques of X-ray. However, arguments were raised that this field was not being able to provide explanations for sensations such as hunger, appetite and nausea, or the precise causation of abdominal pain and discomfort.[105] Overall, there existed a complex interaction between the various types of medical disciplines arguing for the validity of their methods.

The milk diet invented by William Brinton still proved highly popular, although physicians strove to find alternatives to it. Yet the phrase 'a milk diet or your life' persisted as a common warning given to patients by their physicians.[106] An alternative to this therapeutic regime existed in treatment by acids and alkalis. However, despite the interest that physiologists had held in gastric acids, there persisted great differences of opinion regarding the quantities of acid necessary. Medical authors debated questions such as the quantities needed and whether stomach acids might act as a stimulant or disinfectant.[107] One popular therapeutic regime that was developed during the early twentieth century was the Sippy diet. This rested upon physiological concepts which stipulated that the movements of the pylorus were largely controlled by the presence of acids in the stomach or duodenum. Peptic ulcer was thought to result from excesses of acid secretion. Bertram Welton Sippy (1866–1924), the American physician who developed the diet, argued that by removing, or at least inhibiting, the actions of harmful acidic fluids through the administration of alkalis, healing would occur, and that the alkalis would act as direct sedatives to the stomach wall. He also observed that gastric acids were produced more vigorously in the late afternoon, meaning that it was preferable to control acids at that time.

The diet certainly appeared to relieve distressing symptoms which might have burdened the patient for months or years. Pain would disappear on the second or third day, and vomiting, flatulence and eructations would cease. Appetite also improved which allowed the patient to regain weight. Furthermore, the mental outlook of the ulcer victim appeared to improve. It was thought that ulcers could completely heal after around five or six weeks of the consumption of three ounces of milk on a hourly basis.[108] British physicians made enthusiastic attempts to popularize this method, not least because its scientific grounding helped them to counter the perceived intrusion of the surgical community into traditional areas of their work. By 1928, it was in use by many staff and recent graduates at St Thomas's Hospital, for instance, and it appeared that patients had little difficulty with the treatment, due to it being simple, requiring little supervision and could even be carried out at the patient's home, however humble his or her circumstances might be.[109]

A further key area where physicians attempted to carve a therapeutic niche for themselves was after-treatment. Complaints often focused upon the slackening of the surgeon's interest once the wound had healed and that 'it then falls to the lot of the physician to watch the slow decay of the vital powers which can be, and have been, operated upon'.[110] A. R. Parsons went on to argue in the *Dublin Journal of Medical Science* that the history of the patient treated surgically revealed much subsequent pain, vomiting and gastric discomfort. For him, recovery from operation did not necessarily imply a complete restoration to health as careful dieting and medicinal treatment were required for up to two years despite surgery's claims to full and immediate relief.[111]

Robert Saundby asserted the importance of the role of the physician here, stressing that the health of patients risked deteriorating following an operation if the medical man did not intervene and provide information on suitable dietary regimes.[112] George Herschell, author of *The Surgical Treatment of Duodenal Ulcer* (1910), a text designed to disqualify the validity of operative intervention, stated that it is 'quite erroneous to suppose that there is something magical in an operation which can remove the constitutional condition which allows the formation of an ulcer' and that operations prevented the possibility of distension and diminished the acidity of gastric juice, but did little else overall. As these effects could be produced entirely by medical methods there seemed to him to exist no reason why the patient should be subjected to a potentially life-threatening operation.[113]

The validity of surgical intervention was also undermined by discoveries that some key operations were ineffective, and had in fact created a new spectrum of internal problems. Confidence in surgical ability diminished as physicians began to consult patients supposedly surgically cured but who had returned to the clinic with recurrent surgical problems. Gastro-jejunal ulcer, for instance, was a new, artificial disease that only occurred once a stomach operation had taken place. Abdominal surgery itself had produced a new type of ulcer.[114] The procedure of gastroenterostomy also fell into disrepute to such an extent that in 1923, William Hugh Cowie Romanis, surgeon at St Thomas's Hospital, London, looked back retrospectively on the history of the operation and concluded that it was 'the most dismal operation in surgery'. He argued that the profession had been so impressed initially by its uses and the soundness of its scientific basis that it was vaunted as a cure-all for nearly all gastric and duodenal problems. However, the procedure fell into disrepute after a few years and, after seven or eight years, Romanis explained,

> it was distressingly frequent to find patients presenting themselves in the out-patient department of any big general hospital who had had their gastroenterostomies done all over the country ... quite impartially, and who now came up complaining of the same old pain and symptoms, sometimes even worse than before the operation.[115]

Leading gastroenterologist Arthur Hurst (1879–1944) went so far as to claim that 'everyone must agree that the surgical treatment of gastric and duodenal ulcer is a confession of failure'.[116]

Surgeons were quick to defend their position. Moynihan argued that the necessity of their intervention in gastric ulcer was, conversely, a confession that medical treatment was failing to provide permanent healing, although it did relieve pain and curtail attacks of gastric ulcers. For him, the surgeon's ability to visualize ulcers had made them more aware of

> how considerable an event it must be for sound healing to take place, and how prone the large and deep scar which so results may be to break down under a provocation that the normal stomach would perhaps easily resist.

He claimed that medical treatment was haphazard and perfunctory as hospitals could not afford to keep patients demanding long attention.[117] Moynihan also criticized the Sippy method for being carried to such extremes that the oral administration of alkalis was harmful, suggesting that its theoretical basis was problematic as the diagnosis of gastric ulcer was still so vague that attempts to neutralize acids were futile. For him, diagnostic certainty in such conditions existed only on the operating table, with medical treatment potentially being given for ulcers when patients might be suffering from an entirely different disease.[118]

One factor that worked in the favour of surgeons was the patient's economic conditions, particularly during the financial crises of the 1930s. Recurrent periods of disability heightened anxiety about employment security as the medical treatment of gastric ulcer required lengthy periods of time away from work. Surgical treatment might appear preferable as it offered a quicker, more permanent method of relief.[119] Patients felt that it was of great importance to know what their expectations of working life would be afterwards, and under which form of treatment they would be the best wage-earners, and suffer from less invalidism. As one patient explained to his physician, 'if I could spend six weeks in bed every six months I should be all right, but I can't afford it'.[120]

Yet the striking confidence placed upon the potential of the 'pathology of the living' to revolutionize medicine became rapidly undermined by the increasing success of radiographers to utilize X-ray technology to observe inner organs such as the stomach and duodenum in a far less intrusive manner.[121] The ability of radiographers to investigate the abdomen reached a peak in the second decade of the twentieth century, as it had proven difficult in earlier periods to distinguish between the abdominal organs or to recognize and distinguish between their diseases to a sufficient level. In fact, whether or not high-quality X-rays could be produced of the abdomens of anyone but the thinnest patients seemed to be a matter of speculation until publications such as Alfred Ernest Barclay's *The Stomach and Oesophagus: A Radiographic Study* (1913) began to be published.[122] One problem had been the wariness of administering the large amounts of bismuth necessary to produce a suitable image in case the patient was accidentally poisoned.[123]

It is worth noting that attempts were made to unify the divided fields of medicine. A series of publications were produced with the intention of showing how physicians and surgeons might collaborate. W. C. Bosanquet and H. S. Clogg published theirs on the stomach, intestine and pancreas in 1909. Within it, they claimed that the dividing line between the two professions was 'purely arbitrary' and that 'in no department of medicine is their ready co-operation no more necessary than in that of diseases of the digestive tract'. The authors were particularly careful to recognize that lives had been sacrificed in the past by a failure of the physician to use the surgeon's skill at an early period, while operations had indeed been performed unnecessarily for essentially medical conditions.[124]

Robert Saundby also attempted to resolve the differences between physician and surgeon, hoping to reach a general agreement. He believed it to be crucial that physicians practising as abdominal specialists should come to definite agreement about grounds for operative treatment, believing there to exist disadvantages from the increasing division of medical practice between surgery and medicine. He conceded that British physicians were wrong to dismiss modern surgical procedures entirely, but insisted that they should improve their diagnostic and therapeutic skills if they wished to withhold their traditional role in managing stomach complaints.[125] Overall, he called for the restraint of surgical intervention, 'for brilliant as are its results in suitable cases, surgeons are apt to cut the Gordian knot of diagnosis by an exploratory incision'. It was the non-operating physician, he suggested, who should correctly determine which conditions justified surgical intervention.[126]

Saundby suggested that stomach diseases were in fact common ground upon which the physician and surgeon should meet, and that it was the duty of the practitioner to consult with the latter.[127] In 1905, speaking at a discussion dominated by surgeons, he stated that 'the man who shines as an operator is seldom patient of the slow processes by which a medical diagnosis is reached, and prefers to cut them short by an exploratory incision'. For him, the methodology of the trained physician:

> Appeals to a different but not an inferior order of mind and by no means deserves to be neglected, for it fills a gap which would otherwise yawn very widely. When all the crooked ways of pathology shall have been made straight and all the rough places of diagnosis plain, then may not surgery itself be reduced to mechanism and the surgeon be superseded by a girl pressing a button. Until that day arrives there should be room for both types of consultant.

For Saundby, it was unsatisfactory that a patient who had been advised by a physician not to undergo an operation should afterwards be seen by a surgeon who would pursue a fatal intervention. Such errors, he believed, might be prevented if the two consultants were allowed to meet and discuss the question of treatment, as he himself had done when dealing with patients suffering from haemorrhage from gastric ulcer. Saundby observed that he had witnessed fatal cases only rarely, and was wary of claims that mortality rates had diminished to the extent that operative interference on ulcers should be regularly undertaken. 'This is a condition which is rarely fatal when diagnosed and treated medically', he explained,

> and so long as the rate of surgical mortality was high, the risk to life might be allowed to outweigh the advantage of operation, but at present the operation in skilled hands is one which deserves to receive the support of the profession.

He was particularly opposed to operative treatment for chronic dyspepsia, particularly when of nervous origin, as these patients were typically made worse by surgery and it exposed them to unnecessary risk to life.[128]

Yet despite some attempts at co-operation, the overall view that emerged was that there existed too many varieties of treatment, none of which had particularly valid claims to superior forms of therapy. Tellingly, the *British Medical Journal* angrily argued in 1920 that:

> It is none the less a confession of failure that surgical treatment should be required at all – failure on the part of the physician who has not effected a quick, sure, and permanent cure by medical means; the practitioner who has not recognized the premonitory signs of ulcer and so prevented its development; the medical schools which have not taught their students how these signs can be recognized; the pathologists who have been too much interested in the pathology of the dead and of the lower animals to investigate the pathology of ulcer in the living man; and the Government which has not provided the money to staff and equip institutions.[129]

Conclusions

Duodenal ulcer disease became notably central to surgical discourse following its closer elucidation by abdominal surgeons. It emerged as a flagship disease intended to exemplify the ability which surgeons such as Berkeley Moynihan wished to accrue not only in precise diagnosis, but also in medical research. This formed part of their wider agenda to enhance their professionality and respectability within medicine more generally. The relative vagueness of knowledge of the stomach, and in particular of duodenal ulcer, provided them with a suitable area of conjecture.

Yet by this time, so many disciplines were competing for authority within the management of stomach complaints that early twentieth-century medical literature became overwhelmed with contributions contesting and disagreeing on the matter. This was also reflective of wider problems of acceptance faced by those espousing reductionist models of the stomach. The stomach was a bodily area which had traditionally been a central target of the work of the physician, not least because its management had long been grounded in providing common sense advice based upon dietary change and lifestyle adjustment. The vigorous debate which ensued throughout the early twentieth century represented anxiety over far wider changes taking place within medicine. Hence, discussion of duodenal ulcer disease was not just about technical details surrounding treatment. Opposition to surgical operation acted as a check on the ambitions of new disciplines, and represented an assertion of the validity of medical disciplines.

5 THE PSYCHOSOMATIC STOMACH: BRITISH SOCIETY, WARTIME DYSPEPSIA AND THE RETURN OF THE PATIENT, *C.* 1920–45[1]

Pathological anatomy, physiological enquiry and abdominal surgery had all presented increasingly reductionist models of the stomach, focusing less and less upon the organ as a whole and its relationship with the bodily system, and instead concentrating more intensely on its various constituent parts. Gastric illness came to occupy a central space within the discourses that emerged between competing medical disciplines. An array of procedures and technologies had been developed in line with the reductionist approach to provide for the scientific management of gastric complaints. Yet the nature and application of these contrasted sharply with earlier methods of investigation which had focused intensely upon the patient; emphasized the relationship between stomach and mind; provided space for discussion of a highly interactive internal constitution; and which had often prioritized readjustment to external environments.[2]

Advocates of laboratory medicine, for instance, had expressed a marked enthusiasm for reducing interpretations of disease causation to factors which demanded close localized attention only. High levels of acidity in the stomach, for example, were positioned as being solely to blame for the production of gastric and duodenal ulcers. In terms of treatment, abdominal surgery became concerned simply with removing any anomalies found to provide relief, a strategy that typically provoked minimal enquiry into the more general condition of the patient. Overall, the function of the stomach had been radically reassessed and a tendency had arisen to depict it as holding significantly less corporeal importance than it had been accorded historically.

Yet the reductionist approach proved unsatisfactory for many elements of the medical profession. Furthermore, neither pathological examination, laboratory techniques nor abdominal surgery had truly mastered the diagnostic complexities of the organ, or provided an adequate therapeutic option that satisfied all. This final chapter explores the place of the stomach within British medicine during an era which witnessed a rethinking of the direction that modern medicine had taken, and examines how a reassertion of the role of holistic medicine impacted upon approaches to the medical complaints of the stomach. This facilitated a wider incorporation of concepts derived from psychiatry and psychology, and engaged with a renewal of interest in the wider social and cultural environments which might stimulate the onset of gastric illness. In Britain, the Second World War acted as a turning point in the popularization of these approaches. Perceived connections between a sharp rise in incidence of gastric illness and the psychological impact of war upon British soldiers and civilians accelerated a reassessment of the management of stomach complaints. Ulcer disease became increasingly conceptualized as an index of anxiety, once it began to provide objective criteria for managing the problems of civilian and military morale during the Second World War.[3] This helped to widen the backlash against reductionist treatment of gastric illness and cemented suggestions that a return to a patient-orientated approach to gastric illness was necessary.[4] It also extended the role of psychiatrists within this area of clinical activity.

Holistic Medicine and the Stomach

The existence of interconnections between the major organs fitted uneasily within reductionist models of medicine. Emergent scientific disciplines could not easily explain such interactions by focusing so closely on component parts of organs.[5] Yet an undercurrent of thinking persisted throughout the early twentieth century which lamented the loss of traditional ways of knowing the stomach, and which stressed the potentially detrimental consequences of this upon diagnosis and treatment. For instance, in 1910, William Soltau Fenwick (1881–1961), surgeon at Charing Cross Hospital, London, expressed alarm over the huge discrepancies which he observed to exist between what the stomach was expected to do according to modern physiological and surgical findings, and what it actually did in reality. For Fenwick, drugs intended to control gastric secretion and food items developed to have immediate absorption were found in practice 'to possess one inherent drawback – they seldom do what is expected of them'. Unsurprisingly, given his views on the matter, Fenwick was eager to assert traditional concepts that stressed the sympathetic affections of the nervous system and the role of mental influences. He was quick to dismiss physiological research on the basis that there existed obvious differences between the body of the laboratory dog and the dyspeptic human. Chemical analysis, Fenwick maintained, could never explain clinical phenomena such as patients sustaining physical shock, violent emotions or depressing news, and then developing severe forms of gastric illness.[6]

Fenwick concluded that although a great amount of ingenuity had been expended attempting to assign the symptoms of various stomach conditions to their exact causes, a purely symptomatic nomenclature had actually 'held almost undisputed sway through many centuries'. He went on to argue that 'many ancient empirical methods are still of the greatest value, despite the fact that experiments are supposed to have proved them to be unscientific in origin and useless in application'. For Fenwick, the stomach was far too complicated for experimental evidence ever to explain the organ's diseases fully. With a firm nod to what he considered to be the more practical medicine of the early nineteenth century, he stated that 'it would appear that Abernethy foresaw the future influence of theoretical teaching upon medical practice when he uttered the well-known words, "The stomach is neither a stew-pan nor a test-tube, but a stomach"'.[7]

Such opinions gained increasing validity. During the interwar years of the twentieth century, just as reductionist medicine was claiming its greatest triumphs, condemnations of the modern direction of medicine became more prominent, and often stemmed from individuals within mainstream medicine. This opposition took various forms including constitutionalism, psychosomatic medicine and social medicine, and can be broadly characterized as 'holistic'. Adherents emphasized, for instance, the importance of the patient as a whole. While relatively few holists challenged the central role of science, they felt that it ought to be supplemented by collective historical and clinical experience, as well as by the intuition which they deemed to be necessary to guide clinical practice.[8]

Given that the relationship between stomach and mind had historically been perceived to be of high significance, it is unsurprising to find that the digestive organs became a central topic of interest amongst those with holistic tendencies, even those working within mainstream medicine. Groups of American physiologists began to express interest in the potential impact of emotion on gastric functions throughout the early twentieth century, for instance. The work of Walter B. Cannon (1871–1945) was especially influential in demonstrating links between fear and the digestive processes. Although never going so far as to pronounce a dominant role for the importance of the mind in the clinical setting when treating gastric disorders, he suggested that the patient's mental state should always be considered in diagnosis and treatment.[9]

Similarly, the American physiologist Walter C. Alvarez (1884–1978) published *Nervous Indigestion* in 1930, a book stressing the need to understand the psychic roots of the individual's digestive distress.[10] Harvard neurosurgeon Harvey Cushing (1869–1939) developed this research, using experimental methods to suggest that peptic ulcer could be produced in animals by stimulation of the hypothalamic region of the brain. His results implied that there existed intricate connections between the irritative lesions located in the brain stem and local erosions in the stomach such as perforating ulcers.[11] Finally, the work of the Hungarian-American psychoanalyst Franz Alexander (1891–1964) encouraged interest in the psychological aspects of ulcer disease, research that had been particularly successful in exploring the dynamic interrelation between mind and body.[12] This was in line with trends in holistic thinking which eagerly incorporated factors such as the emotions and the psyche into the study and care of individuals, and explored the relationship of the patient with their physical and social environments.[13]

In Britain, attempts were made during the 1930s to transfer the results of this international research into the practical context of the clinical setting. Notably, in 1937, Daniel T. Davies, physician to the Royal Free Hospital, London, and A. T. Macbeth Wilson, assistant physician to the Tavistock Clinic, published what was to become a famous paper on gastric conditions which aimed to place the aforementioned research of physiologists and psychologists on a sounder statistical basis. To achieve this, they examined 205 peptic ulcer patients. Their results suggested that certain events causing high levels of anxiety had preceded up to 84 per cent of examined cases. The stressful events identified were wide ranging and included a new job, unemployment, the death of a relative, illness of a partner, 'quarrels with the wife', unwanted pregnancies, diagnosis of illness and even a 'bombing exploit'. Furthermore, out of 52 cases of relapses investigated, 42 were thought to have occurred immediately after a further experience of anxiety.[14] For Davies and Wilson, this seemed to provide firm evidence that peptic ulcer resulted from the ability of mental states to produce structural change within the digestive system.[15]

It was not enough, however, to claim that exposure to anxiety directly resulted in gastric illness. After all, large numbers of people would be exposed to the very same source of anxiety, yet would not necessarily all develop a gastric ulcer. Davies and Wilson therefore applied ideas that people with particular personalities were more prone to reacting to anxiety by developing a severe gastric condition, an idea directly influenced by Franz Alexander. To explain this, the concept of the 'ulcer type' was developed, a descriptive term applied to patients characterized by certain physical and emotional qualities. These were typically male, had a long thin face and slim build, expressed ceaseless energy and restlessness, and had a tendency to suffer from fear or anxiety.[16] Such patients were thought to be able to pass through life happily until they experienced a stressful situation, the emergent tension of which, it was suggested, would be discharged through the channels of the autonomic nervous system and manifest within the digestive organs.[17] Crucially, Davies and Wilson maintained that practitioners should fully acknowledge the influence of emotional upset in causing profound forms of gastric disturbance. It was necessary, they argued, to take into account a far wider range of contributing factors, rather than focusing on the specific lesion alone, including the patient's occupation, responsibilities and social environment.[18]

The mounting influence of concepts which reasserted the intricate relationships between the mind and gastrointestinal tract also encouraged social scientists to make statements about the relationship between the 'ulcer type' and social environments.[19] This stimulated a re-evaluation of the negative effects of modern life and civilization on the stomach's health, in many ways mirroring the viewpoints of many nineteenth-century authors on the problems of the organ. Anxiety became once again presented as an inevitable factor in modern, competitive societies. As believed in the nineteenth-century, it was impossible to abate these shifting socio-economic conditions, meaning that adjustment became an essential facet of proposed therapeutic regimes. Psychiatric approaches appeared to be the most useful method to adjust the patient to such environments, and these, it was argued, would encourage far more positive results than reliance on medical or surgical treatment alone.

This process was also stimulated by broader shifts in British psychiatry, which witnessed a move away from managing severe psychotic and organic conditions such as schizophrenia towards those neurotic conditions and personality disorders which were less likely to lead to chronic disability. This partly resulted from the establishment of non-residential services and the resultant diminishing of Britain's asylum population. Furthermore, psychiatrists became eager to establish their work firmly within the sphere of professional medical activity and to transform perceptions of it as a proper specialism of medicine.[20] The rising influence of psychiatric ideas in the management of gastric disorder was one manner by which this could be achieved, as it aligned their work closer to that of the rest of the medical profession.

The Dyspeptic British Soldier

Medical professionals have often been forced to engage with unexpected problems during times of conflict.[21] During the Second World War, conflict appeared to have generated unexpected changes in the incidence of crippling abdominal complaints, especially dyspepsia and peptic ulcer. Concern over the predicament became particularly pronounced as the government anticipated a resultant drain in national manpower and military efficiency.[22] Abdominal complaints had gone relatively unnoticed during the First World War. Gastritis, for instance, tended only to present itself as a consequence of gassing, the effects of which included abdominal irritation and varying degrees of internal inflammatory reaction.[23] Instead, cardiac problems appeared common within British forces during the First World War, and the phrase 'soldier's heart' became commonly applied as a descriptive term for the phenomenon.[24]

Unexpectedly, however, dyspepsia and gastric ulcers began to affect British troops at seemingly alarming rates almost immediately from the outbreak of the Second World War in September 1939.[25] The predicament, especially in the early stages of the war, became feared to hold catastrophic potential for national military efficiency. It appeared that biological reactions to conflict were coalescing around the abdominal region, causing unexplained, unexpected shifts in levels of digestive complaints. Gastric illness came to be seen as a very real and worsening problem.

Concern over a rising incidence of peptic and duodenal ulcers stemmed from the 1920s, when it was believed that the disease was on the increase at dramatic levels. In 1929, reports had been published claiming that cases of perforated ulcers had risen by over 300 per cent within the male population of Sweden.[26] Perceptions of an apparently mounting incidence became hotly debated internationally. The most influential British study was penned by Deny Jennings, then medical research fellow at the University of Oxford, who argued in the *Lancet* in 1940 that peptic and duodenal ulcers were rapidly increasing in Britain.[27] The publication of Jennings's work was especially timely, coinciding as it did with the apparent exacerbation of the conditions within the military.

Jennings's study had been stimulated by events on the beaches of Dunkirk, where the British military first observed that high levels of ulcer perforations were occurring amongst the troops. Many soldiers who thought that they were suffering from simple indigestion discovered that they were in fact burdened with severe latent ulcer problems which had been worsened by evacuation. In many cases, these ulcers had perforated during the military operation. Reports on the efficacy of treatment vary. One soldier was recorded as having undergone an operation in a French field ambulance concealed within a wood near the town, although, despite the makeshift conditions, he rapidly recovered. Others, however, were not so lucky. Some soldiers were operated upon days after their perforation, suffering from the excruciating pain of peritonitis in the meantime and being carried around on unsuitable transport.[28]

The problem began to be noted in other military scenarios, and a number of studies of so-called 'military dyspepsia' or 'war ulcers' were conducted early on in the war in response. The most influential of these was undertaken in 1940 by Charles Newman and Reginald T. Payne, both from the British Postgraduate Medical School, who were appointed by the Royal College of Physicians to conduct an investigation of military incidence between September 1939 and April 1940. The researchers considered their findings to be so urgent that they published an interim report rather than waiting until they had fully collated their research. Newman and Payne determined that up until April 1940, 14.4 per cent of all medical cases evacuated to the United Kingdom from France had been diagnosed either with gastric or duodenal ulcer. Even this figure was considered by them to be understated as it was probable that there still existed many cases in the French military hospitals.[29]

Further studies undertaken that year seemed to confirm these conclusions. Philip Willcox, medical registrar at St Mary's Hospital, London, examined 260 medical cases sent from France and found that 25 per cent of patients were suffering from acute gastric problems.[30] Medical officer J. Gibson Graham and Captain John Olav Kerr, meanwhile, also investigated the problem within the Royal Navy, Army, Air Force and Women's Auxiliary Service, although army patients seemed to constitute the majority of the cases discovered. While they did not identify a rise in incidence as such, their report raised the pertinent point that it was likely that the problem would worsen as intake of men into the army increased. Graham and Kerr also found that between April and October 1940, 36 per cent of 980 medical men in one hospital had been admitted for radiological investigation of their digestive complaints.[31] Similarly, C. A. Hinds-Howell, a medical specialist working with the Royal Army Medical Corps, determined that in one military hospital during 1940, 270 patients had been diagnosed with disorders of the stomach, a total of 14.6 per cent of all patients, and 52.7 per cent of these were diagnosed as suffering from peptic ulcer.[32]

A number of factors were postulated as being the likely cause of these rising levels of military dyspepsia. Initially, physical factors were prioritized including poor army diet and high levels of tobacco usage. Tellingly, Newman and Payne's investigations placed special emphasis upon the inadequacy of military food, denying altogether possibilities that psychological factors might be contributing in some way to wartime gastric illness. They deemed army food to be so awful that even those soldiers who had lived healthily for years in civilian life on a careful diet appeared unable to cope with a sudden shift to consumption of the heavy, fatty foods available to them. To worsen matters, the constant physical fitness and exercise that was so central to military life encouraged larger appetites. This meant that soldiers would be inclined to consume larger quantities of the deficient food on offer to them, thereby increasing their likelihood of developing gastric dysfunction. Accordingly, Newman and Payne's recommendations emphasized the necessity of the provision of unique feeding facilities for soldiers who had already succumbed to dyspepsia, as well as the improvement of army food. This emphasis on physiological factors alone facilitated a relatively simple solution. All that needed to be done to resolve the dilemma was to improve army cookery. Simultaneously, those with a weak digestion would be quickly weeded out in the early stages of the war.[33] However, in reality, and somewhat unexpectedly, although army cookery improved, military stomach problems continued to be reported to be increasing.[34]

A further postulated cause was high levels of tobacco smoking within the forces. Scientific attempts to forge linkage between smoking and gastric problems had taken place throughout the 1930s.[35] Wilcox's report had placed particular emphasis on the potential role of tobacco after having observed that there were only three non-smokers amongst his many gastric patients. To emphasize his argument, he pointed to the fact that one of his duodenal ulcer patients smoked up to sixty cigarettes per day, a habit which he believed was being encouraged by the provision of reduced-cost tobacco within the Services.[36] Overall, however, specific connections between tobacco and wartime dyspepsia proved difficult to determine fully. Contemporary medical authors quickly disengaged themselves from discussion of it as a primary causative factor.

Psychological explanations became popularized primarily in response to events in the civilian context. Furthermore, the apparent extension of digestive complaints into civilian life did much to intensify fears regarding the gravity of the problem to the nation's overall war effort. The most puzzling occurrences of rising perforated ulcer levels were closely connected to air raids, although the extent to which these so-called 'air-raid ulcers' were real or imagined remained a contentious issue throughout the war. Connections between air raids and ill health had been discussed immediately from the outbreak of war. Initially, explanation had been sought in psychiatric factors. Writing in October 1939, George Pegge, psychiatrist to the Emergency Medical Service, made claims that the anticipated mass bombardment of civilian populations was producing widespread anxiety. In some cases, the sound of sirens was reported to have acted as a direct stimulus to mental breakdown.[37] By September 1940, the London population had become more accustomed to sirens. Yet observations continued to be made of unexpected correlations between neurotic cases and air raids.[38] Unlike in the previous year, people suffering from wartime anxiety might well have been residing in houses that were rocked or wrecked by high explosive bombs, triggering a stronger psychological response.[39]

Yet mental problems were not the only consequence of exposure to air-raid stress. Shortly after the beginning of the second week of the London air raids of September 1940, seven patients were admitted to Charing Cross Hospital with perforated peptic ulcers, a hospital which normally admitted only one per month. Even that low figure had been expected to drop given that London's population levels had decreased rapidly following the city's evacuation. D. N. Stewart and D. M. de R. Winser, two students at Charing Cross Hospital, decided to approach eighteen other London hospitals to see if this increase was coincidental. They determined that the monthly average in London since 1937 had been around twenty-five cases. Yet in two months alone during 1940, this figure had risen to sixty-four.[40] Further investigations into air-raid ulcers between September 1940 and May 1941 seemed to confirm a sharp increase in the condition in those areas subject to bombings.[41]

The investigations undertaken by the two students initially identified the most likely aetiologies as being acidity, vascular changes, miscellaneous drugs, trauma, bacteria and anxiety. Yet, to the researchers, there seemed to be little reason why overactive acidity should have been particularly high in the months of September and October, or any reason why an idiopathic alteration in gastric or vascular function might have suddenly occurred throughout the community. It also seemed highly improbable to them that the use of medicinal drugs which upset the digestive system should have risen so dramatically, with the exceptions of alcohol and nicotine. If infection was a cause, meanwhile, then it was one that the close contact of shelter life would have inevitably encouraged. Furthermore, it seemed to the investigators that the country's health was particularly good at that time, with perforated ulcer incidence having dropped in November and December, a period when infection within the shelters was still a potential problem. It was therefore communal anxiety which provided the most convincing explanatory rationale. Gastric illness became quickly posited as a psycho-physiological response to air raids.[42]

Hans Selye (1907–82), the famed endocrinologist who did much to develop concepts of biological stress in this period, added to the debate by noting that reactions to air raids followed a similar pattern to that shown in his experiments on laboratory animals, upon whom he had produced acute gastric ulcers by means of emotional excitement, toxic drugs, trauma and exhausting forced muscular exercise. Tellingly, he stated in the *Lancet* that 'it may be of interest in connexion with the gastric ulcers seen during air-raids that starvation and cold greatly increase the ease with which such lesions are produced in animals by exposure to stress'.[43]

Reductionist and Holistic Approaches

The significance of these reports lies in the wider context within which they were produced, and the complexities which they added to the ongoing discourses in existence between different medical disciplines on how to manage the stomach. The experience of 'air-raid ulcers' completely undermined the validity of concepts attributing blame to factors such as diet and tobacco, encouraging alternative interpretations which prioritized the potential role of wartime anxiety. When perceived in this context, the mystery of an increase in perforated ulcers at Dunkirk now appeared to be resolved. Anxiety became posited as an interlinking factor shared by those operating within a variety of civilian and military scenarios. This intensified fears relating to the effects of total war. Conflict seemed now to hold the ability to affect even those areas far away from the initial sites of battle. One impact of this was its potential to exacerbate latent disease, revealing weaknesses in the structural make-up of British society.[44] This opened up possibilities for those interested in advancing the cause of holism. Responses to wartime gastric complaints also reinforced suggestions that reductionist models of medicine that concentrated intensely upon lesions alone were failing to take fully into account the wider impact of environmental and constitutional factors in the production of gastric disorders.

Ideas that over a decade of underlying mental anxiety, encouraged by economic depression, had resulted in widespread nervous strain within Britain seemed to have been proven by a communal breakdown in abdominal health. These concepts gained currency as an explanatory tool for the increase in gastric problems as the Second World War was a period of intense social stress. Such theories also answered problematic questions brought up within Payne and Newman's investigations of 1940, which had identified that the first symptoms of almost 90 per cent of the stomach complaints under their investigation had developed in the 1930s.[45]

Once potential psychological dimensions became established as a likely cause, it seemed logical to assume that unexpected epidemiological shifts in fact resulted from some underlying factor shared by a high proportion of members of British society in the preceding decade. Economic depression, it was suggested, had created internal problems which had mostly remained latent until wartime anxiety had provoked them, and worsened their severity. Socio-economic events of the preceding decade appeared to have laid the groundwork for a sudden manifestation of widespread, debilitating abdominal conditions. The weakened stomachs of soldiers and civilians became conceived as an index of the effects of years of economic hardship and social anxiety.

Psychological theory was easily applied to explain the predicament which the British military found themselves in from 1939. The sudden increase in severe gastric conditions was as a direct consequence of the entire country suddenly being placed in a state of nervous and psychological strain. It was noted that from the summer of 1940 onwards, there had been a general increase in overtime work, as well as additional duties for those still employed in their usual occupation. The resulting weariness, lack of sleep, and irregularity of meal hours were persistently considered to be a major contributing factor to the increase in perforating ulcers.[46] Although dietary changes, hurried and irregular meals, smoking and a lack of fresh air in blacked-out bedrooms were deemed as unhelpful, a lack of sleep, worrying about finances or family members and even anxiety over the distressed state of mankind were all hypothesized as potentially causative factors of a prevalence of problems of the gastrointestinal tract.[47] The apparent susceptibility of the male population of London to them could be easily attributed to the higher frequency of night-time air-raid duties in addition to regular daytime jobs.[48]

It also seemed that the 'ulcer type' could be located in both civilian and military life. In the army, medical authors often identified him as the over-conscientious non-commissioned officer, or a driver of a motor lorry, a job entailing prolonged, strained attention and often leading to underlying feelings of anxiety.

At home, he might be employed as a busy clerk working long office hours in a responsible position, smoking innumerable cigarettes and eating meals at long or irregular intervals.[49] Such explanations were easily applicable to the condition of the dyspeptic soldier, discussion about whom began to focus upon issues such as financial worry, anxiety about their return to civilian life and distress about whether their families might be suffering from bombings back at home.[50]

These concepts, although they gained credibility throughout the war, were most fully developed in 1944 by the pioneering social researcher Richard Titmuss (1907–73), who composed a detailed statistical investigation of the problem with Lieutenant-Colonel Jerry Morris (1910–2009) of the Royal Army Medical Corps. They concluded that the economic patterns of unemployment and re-employment witnessed in the 1930s were in fact reflected in the epidemiological behaviour of peptic ulcer disease in that period. At the start of the decade unemployment seemed to have led to a reduction in ulcer mortality as the death rate from the complaint had dropped in those areas particularly affected. However, when unemployment declined as economic depression eased, ulcer mortality seemed to have risen sharply. It appeared to be the return to work that encouraged higher death rates. Titmuss and Morris were also highly concerned with the rise of city populations during the inter-war period as, if their observations were correct, such areas were most likely to attract the restless, energetic and ambitious 'ulcer types'. Overall, they made recommendations that the medical profession should pay close attention to new nervous strains and stresses that might contribute to the increase in gastric disorders, especially in urban and industrial life, and to utilize these observations in order to help them manage such problems. Clinico-social investigation should, in their view, pay closer attention to the interplay of constitutional and environmental forces.[51]

Gastric Therapeutics and the War Effort

These interpretations of the causative factors of rising dyspepsia and peptic ulcer levels helped to further challenges being made to reductionist medicine, not least because they seemed to underline the often forgotten importance of constitutional factors and the formative role of socio-cultural environments in the production of gastric illness. This held important implications for the ensuing debates regarding the management of the stomach. It seemed to an increasing number of medical professionals that it was necessary for therapeutic interest to move beyond analysis of the contents of the patient's stomach and the speedy surgical removal of medical anomalies, and to recognize ulcer disease and dyspepsia instead as conditions with multiple causes and cures.[52]

The utility of psychiatric ideas in the management of gastric illness became more fully realized and applied as a direct result of the experience of wartime dyspepsia. Furthermore, the potential arena of influence of those with particular skill in matters relating to digestive health became enhanced. The British military was particularly willing to engage with a wide range of medical specialities throughout the war, and it was those advocating holistic views who gained the greatest authority in this area, at least in the treatment of gastric illness.[53] As early as 1942, the manpower situation in Britain appeared to be becoming so acute that the country could not afford to operate with a disregard for causalities in the same way that larger countries such as Germany and Russia could.[54] Medicine therefore came to play a central role in manpower conservation as it would help to relieve casualties. It is in this context that close relationships were forged between medical and combatant officers. If the latter were medically minded, then great military advantages might be conferred.[55]

New roles for the gastric specialist emerged in both civilian and military contexts. At home, members of the medical community became highly vocal with their opinions on how efficiency at home could best be maintained. They were often concerned with disseminating knowledge on how best to keep the gastric patient busy at work. For instance, in 1943, the physician J. B. Wrathall Rowe wrote to the Ministry of Labour suggesting that the rationing of food supplies such as milk made treatment difficult for patients who had been recommended particular diets. Rowe provides an interesting example of a stomach ulcer sufferer who was both patient and doctor during the war, and he was able to claim from personal experience that he had faced extreme difficulties obtaining the full, varied diet necessary for recovery.[56] Gastric expertise was also called upon when special dietary provisions were constructed for factory workers. This also helped to maintain national manpower, although the extent to which these were taken up by employers appears to have varied. Tellingly, in 1944, the London physician J. J. Horwich wrote to the *British Medical Journal* claiming that he had recently had a discussion with a medical officer working at a large local factory after a patient had told him that he could not get any special food at his work canteen. He replied that in his factory ulcer cases were allowed special diet, but only six out of 2,000 employees had taken advantage of the light food available, despite prominent notices in the canteen.[57]

However, the influence of gastric specialism proved to be far stronger in the military. Tellingly, those with expertise in the physiology of the digestive tract were called upon to provide guidance in the enlisting of Royal Air Force pilots, a group considered as particularly prone to ulcer complaints due to the apparent relationship between aviation and gastric problems. Aviation had long been observed to have unusual effects upon the digestive tract. Knowledge on matters relating to altitude had been developed in the 1920s and 1930s through experiments with low-pressure chambers, the results of which had suggested that a lack of oxygen modified the gastric functions through the automatic nervous system.[58]

The consequences of this might be potentially disastrous in military scenarios. The perforation of an ulcer whilst flying could result in the escaping of the stomach's contents into the abdominal cavity, resulting in sudden, excruciating pain which made safe flying and landing difficult, if not impossible. The first recorded case of so-called 'abdominal emergency whilst flying' occurred in June 1930, when a thirty-year-old RAF corporal based at Reykjavik developed acute abdominal pain whilst flying at an altitude of 2,000 feet. From this time onwards, fears were heightened that aviation might cause the perforation of latent gastric ulcers.[59] Those with medical expertise of the gastric system also made recommendations for enrolment, suggesting that both the psychological and gastric health of potential recruits should be closely scrutinized. This seemed particularly important when recruiting pilots, as if physical unsuitability for air service was recognized early on, then time and money would not be wasted on lengthy, expensive training.[60]

Dismissal was a further area where prevalent medical opinions on gastric problems became utilized. The leading British gastroenterologist Arthur Hurst became a highly vocal figure on this subject. Hurst is considered to have started gastroenterology as a specific field within British medicine, having founded the Gastroenterological Club in 1937. Importantly, he had become highly interested in the interaction between psychoneuroses and gastric problems.[61] In September 1939, according to Hurst, 50 per cent of patients crippled with gastric illness had been sent back to the Front, yet this had been reduced to 29 per cent in October, and then only 7 per cent over the following eighteen months. Hurst suggested that more care should be taken with military admission and dismissal, advising that no-one with evidence of ever having had an ulcer should be accepted for service. Even if the patient appeared to be free from his condition, the psychological nature of the complaint, he claimed, meant that the ulcer diathesis would remain for life, ensuring that he was liable to recurrences.[62]

Dismissal became an increasingly problematic option as the apparent extent of military stomach complaints became clear. In fact, it was to become such an issue that even as early as 1940, army authorities felt it necessary to send guidelines to the presidents of various military medical boards recommending that chronic dyspeptic cases be transferred to lower classes of service rather than sent back to civilian life in order to maintain military efficiency.[63] If dismissal became increasingly perceived as an unfeasible solution, then an alternative suggestion put forward was to establish 'ulcer battalions'. These were to consist of groups of soldiers placed together to form light duties. It was suggested that the army could arrange special diets and mealtime routines to decrease sickness levels, and it was also recommended that those with gastric expertise would best be able to manage such a programme.[64] This scheme was even claimed to be preferable to the 'ulcer type' as these typically had the kind of personality that would be most likely to object to discharge.[65]

Ultimately, the British army did not take up the idea of ulcer battalions. However, it became increasingly common for soldiers to be given reduced work levels and to remain in normal service. If their condition failed to improve, it was only then that they might be discharged. Tellingly, one soldier whose health deteriorated after Dunkirk was eventually diagnosed with a duodenal ulcer, and allocated as eligible for shore service only. He acted as house help to a lieutenant commander and his wife, with his military duties now typically involving cooking and keeping the house tidy. His health problems continued to deteriorate, and required further hospital treatment in October 1943. It was only when he had been given increasingly reduced levels of service that he was discharged.[66]

Yet if soldiers were expected to continue with their duties, then this necessitated the provision of adequate forms of management of their conditions whilst in service. Suggestions forwarded were often those intended to deal with general aspects of the patient's problem, rather than those dependent upon models derived from reductionist medicine. It is here that the dismissal of a prioritization of physiological therapy can be seen being more fully rejected in favour of greater emphasis on its psychological aspects. By the Second World War, the British military were highly interested in the psychological make-up and well-being of their soldiers.[67] Physicians had suggested that soldiers could maintain a robust digestive system by carrying small packets of drugs around in their pockets, so that they had access to medication at all times.[68] Members of the surgical community, meanwhile, made claims that their forms of therapy were especially useful, supporting their suggestions that rapid recovery from an operation offered a speedier solution than drawn-out forms of medical treatment.[69] This concept, to members of this discipline, appeared to be highly transferable to the military situation as the rapid return of the soldier to active service was of crucial importance.[70]

An increased emphasis upon the potential impact of anxiety in producing gastric illness encouraged suggestions such as these to be rejected, and stimulated calls for a complete overhaul of gastric therapy. Tellingly, Henry Letheby Tidy (1877–1960), a leading author on matters of medicine in wartime, complained that the elaborate dietary schemes available, mostly based upon complex principles of acid reduction, were unsuitable for the treatment of wartime gastric conditions because patients were now in a scenario where therapeutics required urgency and simplicity. Tidy complained that too much attention was being paid to 'the finer details' and that these methods 'should be swept away in the new treatment of wartime'.[71] Surgeons, too, became criticized for being hasty in their enthusiasm for removing huge parts of the stomach or duodenum, a procedure described by one critic to be 'so absurd as to be incredible'.[72]

An emphasis on psychological treatment seemed to offer great potential for the problematic area of gastric management, not least because it incorporated simpler therapeutic models reliant upon rest and behavioural change. It is therefore unsurprising that British physicians persistently went so far as to claim that the war was a timely occasion to revolutionize the treatment of peptic ulcers. The Liverpool physician Robert Coope, for instance, claimed that wartime drug shortages had in fact 'purged gastro-intestinal therapeutics of the fussiness and faddiness which have been their particular bane'. He went on to state that 'even a shortage of newsprint may have compensations if it discourages the hitherto insistent invitations to self-diagnosis and treatment of indigestion or constipation'. Rather than subjecting soldiers to complex routines of dietetic or drug treatment, he insisted that all that was needed for an ulcer to heal was physical and mental rest involving six weeks in bed, followed by a quiet life and regime to safeguard against relapses.[73]

Overall, the war helped to popularize ideas that gastric conditions could not be treated by a focus on physiological aspects alone, whether or not they were treated surgically or medically. It was increasingly deemed once again necessary to look at rest, environmental readjustment and the possibility of multiple causative factors when deciding upon therapeutic choices, the success of which depended upon the personality of the particular patient. Both gastric specialists and psychiatrists called for a wider acknowledgement of the application of these ideas. The influential psychiatrist Aubrey Lewis (1900–75) claimed that, whilst working at the War Office Selection Boards, he had realized the importance of connections between abdominal complaints and personality types and attempted to develop a series of studies utilizing psycho-diagnostic methods on this basis. He hoped to use his findings to convince practitioners to incorporate psychiatric elements into medical therapeutics, pointing out that rest, sedation and environmental readjustment were far more likely to produce positive results than any of the forms of treatment in general use.[74] It is worth noting, however, that psychiatrists never wished to take over treatment completely. They stressed that psychiatric testing should form a part of the clinical examination, but that this did not necessarily depend upon a psychiatrist being present.[75] In Britain, this formed part of their professional agenda to bring their work more into line with medical practice in order to enhance their authority within the various medical disciplines. Yet what psychological concepts inadvertently achieved was to stimulate a reassessment of how gastric illness might best be managed.

Tellingly, Arthur Hurst believed there to be more gastroenterologists in the army than in civilian practice, yet their expertise was being wasted in routine and administrative military duties. He went so far as to argue that the 'soldier's stomach' was a problem created entirely by bad medical management and delays in treatment rather than being a product of specific conditions in army life. Hurst went on to suggest that one way to make savings on manpower was to establish specialist units staffed by experts with gastroenterological experience. His proposed unit was then to be split into three divisions: one for diagnosis; another for those suffering from organic disease; and a third for those suffering from functional dyspepsia. He anticipated that patients diagnosed as suffering from ulcers would be immediately discharged and transferred to a civilian hospital with the hope that they might prove more useful at home.

Hurst's recommended treatment involved a mixture of medicine and psychotherapy, as it was hoped that the application of the latter would ensure that only a small amount of patients would develop into hopeless cases requiring discharge. It was thought that these hospitals would encourage psychological good, with the time spent there being used to introduce new ways of thinking, imbuing valuable life lessons which could be utilized by the patient for the rest of his life. Hurst described it as 'preliminary life training'. Although he did not entirely deny the importance of providing the patient with information on issues such as the neutralization of stomach acids by foods and drugs, he considered it essential to stress the potential role of worry and anxiety to the patient, and to explain the value of this in upholding the health of the general nervous system through periods of mental and physical rest, congenial surroundings, cheerful companions and appetizing food. Furthermore, the treatment rarely required elaborate, specialist methods meaning that any intelligent and sympathetic medical officer could employ it.[76]

As indicated above, those writing about the prevalence of stomach conditions were also interested in social trends. It is therefore unsurprising that Hurst points to the analysis of wider social patterns that might lead to anxiety and other varieties of mental problems, insisting that the medical community needed to engage with wider social aspects of civilian organization. Rising levels of dyspepsia, ulcers, abdominal cancers and inflammated stomachs could only fall, he suggested, if the patient lived in social conditions free from anxiety. Not only would this automatically reduce smoking and alcohol consumption, but, so Hurst argued, freedom from want would lead to better provision of food whilst widespread education of young women in domestic service should lead to better cooking, improved labour conditions would discourage the bolting of meals, whilst adequate holiday provision would reduce digestive problems caused by work fatigue and anxiety. He also recommended that every town hospital should be provided with a county annex for the treatment of chronic diseases as 'no greater contribution could be made than this to the cause of national health'.[77]

Conclusions

The experience of dyspepsia and ulcer disease during the Second World War provides a further example of how the stomach and its illnesses have occupied prominent spaces within the professional activity of British medical men. It also marks a turning point in the wider story of the organ which this book has addressed, as it reasserted long-standing concepts and medical traditions. It challenged the reductionist approaches to the body embedded within the methods and procedures of abdominal surgery and laboratory medicine. Furthermore, it helped to bring about a more standardized method of treatment, a response to the high levels of contestation which the management of conditions of gastric ulcer had been subjected to. This laid the groundwork for the conceptualization of gastric ulcer as being intrinsically stress-related, a theme that was to dominate discussion of the condition well into the late twentieth century.

CONCLUDING REMARKS

This study has explored the British experience of gastric illness in the broad period 1800 to 1945, addressing gaps in the existing historiography of disease and medicine. It now appears clear that the stomach is an organ that persistently occupied a prominent historical position in the medical sphere as well as within wider socio-cultural arenas. The Victorian preoccupation with digestion and indigestion; the centrality of the stomach to narratives of physiological and surgical medicine; and the prominence of dyspepsia and peptic ulcer within the medical experience of the Second World War have been presented as neglected, but highly significant, topics. In doing so, I have aimed to explore the inter-relationships between various medical disciplines in order to reveal reductionism to be a highly complex and contested entity. Forms of gastric illness have been presented as central to the discourses which emerged, as competing areas of medicine fought for space and professional authority. I have also provided a study which is suggestive that chronic illness is just as worthy of analytical attention as the epidemic diseases regularly discussed by medical historians.

A further observation worth making is that there still exists great research potential in this area of enquiry. The last half of the twentieth century offers ample research opportunities. This book has only touched upon the rise of the stress model within explanations of gastric ulcer, although it is worth noting that this was to have a huge impact upon both medical and popular thinking well into the 1980s.[1] The period also witnessed dramatic therapeutic changes that went some way to undermining perceptions of the necessity of surgical intervention. Notably, James Black (1924–2010) introduced H_2 receptor antagonists in the 1970s which were thought to allow ulcers to heal without recourse to surgery and to remain healed if drug treatment was continued. They worked effectively by blocking the action of histamine on parietal cells in the stomach, thereby decreasing their ability to produce acids. Their impact upon the treatment of problems such as dyspepsia was striking, although they were later to become largely surpassed in popularity by the more effective proton-pump inhibitors. Analysis of this innovation would hold significance for understandings of the relationship between gastric problems and pharmaceutical companies such as

GlaxoSmithKline who became able to amass enormous profits by selling their product under the name Tagamet, which was ultimately to become generally recognized as the first blockbuster drug.[2] Finally, historical analysis of the impact of *H pylori* deserves serious academic attention. The shift towards conceptions that the organ needed to be kept free from infection had a huge impact upon the management of gastric disorders throughout the closing decades of the twentieth century and beyond. It is perhaps no exaggeration to say that the aetiology of peptic ulcer as being bacterial in origin transformed centuries of thinking about gastric illness, not least because it was a completely unexpected aetiological factor.

The Rise and Fall of Peptic Ulcer?

A further important issue which a book on the history of gastric illness cannot avoid is the influential theory of the 'rise and fall of peptic ulcer' between around 1800 and 1950.[3] At present, the leading researcher in this area is Jeremy Hugh Baron, a retired surgeon with gastric expertise at the Mount Sinai Hospital, New York.[4] His research is primarily concerned with the historical activity of *H pylori* bacteria. Initially, the discovery of this as a causative factor of certain forms of gastric illness appeared to have ended centuries of aetiological debates which focused upon factors including acid levels, diet and reactions to anxiety. However it soon became apparent that explaining long-term trends in its occurrence still remained problematic as the presence of the bacteria alone was not sufficient to produce ulceration. It required some elusive form of stimulation by an external agent or condition to encourage it to produce disease.[5] Accordingly, speculation has emerged about how external factors such as industrialization and increased life expectancy might have potentially affected long-term trends in ulcer incidence throughout the nineteenth and twentieth centuries and how the discovery of this factor might impact upon the present-day management of gastric ulcer.[6]

Relying almost exclusively upon statistical evidence accrued from hospital records, Baron has argued that physicians first observed gastric ulcers on a significant scale during the early 1800s, which to him indicates some degree of linkage between the industrial revolution and the behaviour of *H pylori*. It has been hypothesized that peptic ulcer incidence reached a peak in the latter half of the century and then declined at the turn of the twentieth century. Yet this was quickly superseded by a rise in duodenal ulcer, an illness whose incidence rapidly increased until it began to subside in the 1950s. This increase is hypothesized as attributable to either a new pathogenic strain of *H pylori* or a change in host susceptibility. According to the birth-cohort theory often favoured by those interested in this strand of research, increased hygiene standards shifted the age of infection from early to late childhood in the late nineteenth century. This meant that generations born prior to this became infected as toddlers, developing chronic or atrophic pangastritis which stimulated a reduction of the acid secretion that protected them from ulcer disease. Gastric ulcer could only develop in subsequent generations who contracted *H pylori* in childhood and adolescence. Acquisition at an later age resulted in a more limited gastritis with less reduction of acid secretion which has been utilized to explain the rise of duodenal ulcer instead.[7]

The findings presented in this book offer a simpler, and more convincing rationale for perceived shifts in epidemiology. A major point in this book is that shifting forms of management, treatment and diagnosis have ensured that gastric complaints have been regularly reinterpreted, reassessed and often contested even within relatively short periods of time. The professional aspirations of the various medical disciplines as they arose throughout the period of reductionism played a large part in forging understandings of gastric illness. Furthermore, shifts in diagnostic trends have often arisen from changes in professional activity rather than stemming from adjustments in disease behaviour. As this book has shown, diagnostic categories are not static entities and are therefore not, for the most part, coherent with modern diagnostic observations. This is an important point given that authors on the subject of the rise and fall of peptic ulcer have assumed that nineteenth-century medicine brought advances and greater accuracy in diagnostic and technical skills which means that the writings of physicians from the period can be fully relied upon.[8] Certainly, Baron has explicitly stated that hospital diagnostic records had improved to such an extent by the mid-nineteenth century that the statistics of gastric disease produced are inevitably accurate.[9]

However, this presupposes that diagnosis has 'developed' in a linear manner, as medical theory, practice and technologies gradually improved. It does not account for fluctuations or contestation. And it is, of course, contestation and fluctuations in the popularity of ideas regarding gastric illness which this book has emphasized as being a central factor to medical practice. For instance, less tangible factors such as the impact of emotions or psychological states permeate the history of stomach disorders, but their utilization within diagnosis has fluctuated depending upon how easily such concepts could be transferred to reductionist or anti-reductionist trends in medical thinking at any given time.[10] In fact, disease categories can be entirely forgotten about for long periods as diagnoses and therapies go in and out of style. Male menopause provides a striking example of this, with the popularity of its diagnosis depending upon a host of interconnected medical, technological, cultural and economic factors.[11]

Yet epidemiologists have found a way around this problem: simply rediagnose and correct anything which looks inaccurate in the historical record.

This allows disease labels, especially those which look particularly peculiar, to become consonant within a modern medical framework. The obvious problem with this approach is that in order to understand historical diagnoses fully, we need to appreciate and understand the context within which they were produced and the professional motivations which might have encouraged those working within the medical sphere to pay more attention to a complaint, or a region of the body, at any given time.[12] Diagnosis is highly complex from a medico-historical perspective, and it is difficult to imagine that any medical label can be fully understood outside its relevant representational framework.[13] To assume otherwise is simply ahistorical. This approach proves particularly problematic when applied to epidemiological models of gastric illness which stretch over a period of 150 years or so.

A further assumption made is that once a disease entity is identified, then knowledge of it spreads rapidly throughout the medical community. Of course, this is overly simplistic. We cannot assume that just because ulcer of the stomach was first fully described in 1828 by John Abercrombie then all medical practitioners began to utilize it diagnostically, and with accuracy. As the historian Charles F. Wooley stated, 'the translation of clinical observations into conventional clinical wisdom, which in turn is challenged and modified by new technology, is a relentless process – rarely volcanic, all too frequently, glacial'. Advances in research and clinical utilization can often be measured in decades rather than months, argues Wooley, neatly citing the example of blood cholesterol which took nearly a century to transfer itself from its initial measurement into clinical medicine.[14] This was certainly the case with ulcer of the stomach which, as I have shown in Chapter 2, took several decades to become established and accepted as a recognizable disease entity following its emergence in the pathological literature.

An especially notable example of the danger of looking at medical evidence from the nineteenth century without paying close attention to its historical context can be observed in an article written by Baron in 2002 where he cites a statistical increase in the number of publications discussing ulceration as supportive evidence of his general conclusions.[15] If we compare his findings, which imply a steady rise in publications between 1800 and 1930, to the data accrued by William Bynum and Adrian Wilson on the rising numbers of medical journals from 1800 onwards,[16] it becomes apparent that changes in the medical profession at that time prompted a general rise in medical articles, with it being inevitable that articles on a particular medical complaint might have increased at intervals when new discoveries were made, or when innovative methods of treatment were introduced. It is hardly a coincidence that descriptions of stomach ulcers began to emerge at precisely the same time that medical journals started to be produced in Britain in large quantities.[17] For example, advances in stomach surgery from the 1880s onwards led to fresh debate about the efficiency of surgery to treat stomach complaints. An increase in publications is not necessarily a result of an increase in rates of disease. It is necessary to read the articles rather than simply to count them.[18]

A further point to make is that carelessly utilizing statistics to determine the historical behaviour of the bacteria does not allow for possibilities that ulcer disease is inherently multi-causal in nature, as well as being highly complex. Statistics are relevant to developing medical explanations, but correlations themselves have no explanatory force, since they may be the result of confounding alternative causes.[19] The conclusion that there is a causal relation between any particular factor and human disease is highly problematic, not least because explanation cannot be found in terms of one single cause. The production of a gastric ulcer is a complex process involving multiple interacting factors, and is not entirely bacterial.[20] As this study has argued, medical professionals have historically identified a wide range of factors which they believed to encourage the onset of problems such as gastric ulcer. Even in the era of reductionist medicine these views persisted and, indeed, formed a central part of the backlash against it.

A key question to this debate is therefore whether recorded incidence increased as a direct result of observations made by physicians and surgeons of rising levels of gastric ulcer, or whether it changed because medical professionals started to look into new areas of the body. One of the key recurrent themes in this study is the manner by which the stomach became prioritized as an essential bodily organ at the start of the period in question, and how reductionist models of medicine then increasingly segregated and divided the stomach into its component parts, shifting emphasis towards elements such as the duodenum. Abdominal surgery in many ways completed this trend. Hence, it is unsurprising that conditions such as ulcer of the stomach show a rise in recorded incidence during the 1820s and onwards as this was the period when the importance of the stomach within medical literature became especially heightened. Physicians and pathological anatomists simply were not as interested in smaller parts such as the duodenum in the context of prevalent medical paradigms that stressed the significance of organs to medical discovery. Certainly, there appears little to suggest that the activities of these historical actors were particularly influenced by a perceived increase in the disease itself, not least because contemporary investigators believed themselves to be uncovering medical problems that had previously remained unknowable rather than ones that were on the rise. Diagnoses of ulcer of the stomach emerged as a direct consequence of a shift in the professional project of medicine and the ambitions of those working within the field.

This raises the point of why duodenal ulcer was not readily diagnosed up until around 1900. Can we rely entirely on the supposition that as it was not observed, then it simply did not exist to any noticeable degree? Or could it be, perhaps, that contemporaries were so concerned with the stomach as a bodily centre of nervousness, sympathy and disease that the pathological and diagnostic gaze was not directed lower down towards the duodenum? The second proposition appears more probable. Certainly, early nineteenth-century contemporaries were prone to pointing out that there was a distinct lack of knowledge relating to the duodenum despite its potential importance as a 'second stomach'. This was emphasized by a medical author named G. D. Yeats who published *Some Observations on the Duodenum or Second Stomach* in 1820, and who stipulated that virtually nothing had been written of its diseases in the medical texts available. As a result of this lack of attention drawn towards the organ, it seemed that its diseases had been persistently attributed to the liver, and treated as such.[21]

John Abercrombie also noted that clinical information on the diseases of the duodenum was lacking, not least because it was the seat of medical problems which were commonly mistaken for stomach or liver complaints. In addition, he also recognized that ulcer of the duodenum might be fatal by perforation and rapid peritonitis, in a similar manner to that of the stomach.[22] Duodenal ulcer was therefore certainly not absent and, indeed, accounts of it were regularly published in the medical press.[23] However, it did not attract the same attention as ulcer of the stomach because the duodenum was not subject to the same levels of pathological, medical and social attention as the organ. It was also not subject to the same mechanisms of knowledge dissemination. In fact, ulcer of the duodenum was typically only discussed as a matter of interest in relation to burns, as this subject fitted more easily into the predominant interest in concepts of sympathy and bodily interaction. The British surgeon Thomas Blizard Curling (1811–88) was the first to draw attention to this phenomenon which he conceived to be due to some mysterious sympathetic connection between the skin and stomach.[24]

Duodenal ulcer incidence, it has been claimed, underwent a dramatic epidemiological shift around the turn of the twentieth century, which has been described by gastroenterologist Amnon Sonnenberg as 'stupendous' and 'amazing'.[25] Yet it is hard to ignore the rising professional ambitions of abdominal surgeons in this period, and their potential impact upon how the duodenum was observed. As I have shown in Chapter 4, a central part of their agenda to assert the validity of their work was to show how it held fundamental diagnostic potential, not least because this helped to dissociate it from the ethically problematic procedures of laboratory medicine. The early twentieth-century abdominal surgeon claimed his work to be revolutionary, scientifically accurate and increasingly diagnostic in nature and held up duodenal ulcer disease as the flagship complaint which he had successfully identified and managed. This was principally for professional reasons rather than as a response to shifts in disease behaviour.

Certainly, contemporaries did not perceive themselves to be responding to an increasing prominence of duodenal ulcer disease. Rather, they presented themselves as uncovering mistakes previously made in medicine. The alleged failings of nineteenth-century pathology led Moynihan to reject completely post-mortem statistics accumulated in earlier periods. Tellingly, when asked by a critic, the Guy's Hospital physician Arthur Frederick Hertz (1879–1944), if Moynihan believed 'that the bodies of people with duodenal ulcer go straight to another world without passing through the post-mortem room', he replied stating 'most certainly I do believe this of the vast majority. Does any one doubt it? Does Dr Hertz suppose that anything more than a very insignificant proportion of bodies pass through the post mortem room?' For Moynihan, only a small percentage of the total sick in the community had ever been admitted to hospital, and even then, only around 5 per cent of these patients would actually die. Even if the bodies of those were examined via post-mortem, itself constituting a small proportion of the hospital's dead, it seemed to Moynihan that lesions that appeared to be not immediately related to the death of the patient were frequently overlooked. He concluded that 'any inferences formed from the imperfect records of such a small, unrepresentative proportion of the total deaths held little numerical relevance to the diseases of the living'. Accordingly, accepted beliefs on the low commonality of duodenal ulcer disease were based upon 'the imperfect records made in the ancient day when little or nothing was known of the various diseases'.[26]

It therefore seems odd that changes in medical activity should have been entirely excluded from epidemiological studies as a contributing factor for shifting diagnostic patterns, given that they coincided precisely with observations of a dramatic shift in the behaviour of peptic ulcer disease. The relationship between such a sudden alteration in mindset and an apparent rise in duodenal ulcer incidence cannot easily be ignored. In fact, evidence provided in this study based upon firm historical methodology suggests that an increased diagnosis of duodenal ulcer disease at the turn of the twentieth century was not necessarily dependent upon changes in bacteriological or disease behaviour. In fact, strong conditions existed which stimulated changes in medical behaviour and prompted a rapid shift in ideas regarding the diagnosis of a variety of forms of abdominal illness. This is not to say that the techniques of surgery were adopted and approved of by all in the case of gastric ulcer. Yet what is notable is that the diagnostic category which was formulated persisted in popularity, even amongst those opposed to the intrusion of the abdominal surgeon into their traditional area of expertise. The diagnosis of previously vague conditions as new diseases such as duodenal ulcer formed an integral part of the surgeon's redefinition of the problems of the inner body, a factor which had a significant impact upon diagnosis throughout the medical community.

Tellingly, duodenal ulcer became perceived as a disease diagnosed so frequently that even as early as 1911, members of the medical community were warning that the tendency to diagnose it was becoming *too* common. While acknowledging that its existence had been underrated in earlier periods, some contemporaries expressed concern that the tendency was now the opposite; to over-diagnose duodenal ulcer for virtually all gastric complaints.[27] Furthermore, contemporaries rarely perceived the disease as being on the increase, but recognized that certain conditions were prevalent that had encouraged a shift in understandings of a pre-existing medical problem. For instance, in 1913, the *Edinburgh Medical Journal* undertook a collective investigation of the issue, reaching the conclusion that

> we have no reason to suppose that the incidence of duodenal ulcer and its perforation have increased in recent years, and it must be admitted that in all probability some cases passed unrecognised ... no surprise need be felt that such a rare condition was occasionally overlooked at a time when the actual existence of the condition was only beginning to be recognised.[28]

It is therefore difficult to argue convincingly that increases in diagnosis of duodenal ulcer disease should be viewed as separate from the desire of surgeons to revolutionize medicine, to produce modern scientific research of previously hidden diseases and to offer new diagnostic possibilities.

Stress and Gastric Ulcer

As a concluding remark, it is worth noting that the history of the stomach that I have chosen to present throughout this book has much to offer historically minded members of the medical and gastroenterological communities. I have intended it to act as a case study designed to reveal much about fashions in medicine and the impact of these on the clinical experience and, ultimately, upon the patient's encounter with his or her illness. For instance, it has been clearly been shown that factors such as the role of external social environments in the production of gastric illness have been prioritized at certain historical points not necessarily as a direct result of advances or improvements in medical knowledge, as a straightforward narrative of medico-scientific progress might imply. Alternatively, the rise and fall in the popularity of psychosomatic factors have often stemmed from fashion. They might have re-emerged as a response of criticism of its neglect, as outlined within my final chapter, which showed how physiological and surgical models of gastric illness proved unsatisfactory as wartime anxiety clearly appeared to clinicians to be acting as a triggering factor for dyspepsia and ulcer disease.

This predicament rendered the profession open to criticism. Perhaps the most remarkable attack that can be found in the historical literature on the subject of the management of gastric illness was published in 1951 by an obscure author named John Cuthbert Parr, entitled *How I Cured My Duodenal Ulcer*. Within this, Parr details twenty-five years of medical mismanagement beginning towards the end of the 1910s, describing what he perceived to be an alarmingly wide range of medicines administered to him by physicians, all of which had names which were too long and unpronounceable for him to understand. He then goes on to list various operations which he was recommended to undergo throughout his period of illness, including gastroenterostomies and jejunostomies, all of which failed to provide lasting relief. Parr vividly describes his increasing sense of alienation from the medical profession, detailing how his faith in surgery was shaken, and how he entirely lost confidence in the capabilities of doctors to cure. Furthermore, Parr claimed to have met countless other people with comparable problems. Ultimately, he was cured when the physician and surgeon's gaze shifted back towards a holistic view of the body during the 1940s, incorporating a wider range of factors such as anxiety and environmental readjustment.[29]

Although this text is now sixty years old, it has striking resonance for recent claims regarding the rejection of factors such as stress in the production of gastric ulcer disease. Today, the treatment of problems such as peptic and duodenal ulcer disease is principally concerned with the elimination of *H pylori* bacteria since its discovery as a causative factor of ulcers in 1983.[30] Yet this has also led to a rejection, or at least minimalization, of the potential role of psychological dimensions in the production of gastric complaints. At its most extreme, the denunciation of the validity of the mind–stomach relationship has led to claims that by emphasizing psychosomatic factors for decades, attention was drawn away from the development of effective forms of treatment such as antibiotics and models which prioritized bacteria as a potential causative factor, resulting in this being delayed until the 1980s.[31] In fact, the potential relationship between mind and stomach appears to have become so deeply forgotten that in 1998, Michael Gershon managed to present the research contained within *The Second Brain* as an extraordinary discovery which could explain a wide range of internal phenomena including diarrhoea, butterflies, cramps and constipation as well as presenting insights into gastroenteritis, the nervous stomach and irritable bowel syndrome.[32]

More subtle claims have been made that stress is now once again an underrated factor and that the medical profession is currently expressing a tendency to neglect the influence of psycho-social and socio-economic behaviour upon infection rates.[33] Voices in the medical press have persistently attempted to remind readers of the role of psychology in the treatment of gastric ulcers in the mid-twentieth century, largely echoing discussions made throughout the period of the rise of holistic medicine that culminated in a reassertion of the psychosomatic throughout the Second World War.[34] These claims are often backed up by ideas drawn from a line of research which suggests that *H pylori* infection might not be a causal factor in gastric conditions such as dyspepsia, but might in fact be a secondary infection that delays healing.[35]

Notably, research physician Susan Levenstein has recently suggested that the incorporation of psychosomatic factors in the aetiology of peptic ulcer has become *unfashionable* since the increased emphasis on peptic ulcer's potentially bacteriological origin, stressing that there exists strong evidence that psychological stress triggers many ulcers. Bacteria as a mono-causal explanation proves unsatisfactory as many infected people ultimately do not develop ulcers. In her words, '*H pylori* is not enough' despite psychosomatic factors having become virtually stigmatized at the expense of a more dignified explanation of infection. These current debates are clearly not without historical precedent. In fact, they resonate strongly with questions asked at the turn of the previous century regarding an over-reliance on surgical intervention or acid reduction and the loss of the patient within ulcer narratives, a predicament that was ultimately claimed to have detrimental consequences for the patient. Yet the period in which such concepts were reasserted in fact witnessed a flourishing of exciting and varied research opportunities whereby linkage was not only restored but could be reassessed within new frameworks of psychological medicine and neurological research. Could it be too, as Levenstein has recently claimed, that to reconsider peptic ulcer disease as a complaint with multicausal aetiology might also lead to groundbreaking research into the potential relationship between brain and abdomen, breathing new life into psychosomatic concepts of peptic ulcer as well as developing a more general paradigm for applying the integrated biopsychosocial model to medical disorders?[36]

NOTES

Introduction

1. J. R. Warren, 'Unidentified Curved Bacilli on Gastric Epithelium in Active Chronic Gastritis', *Lancet*, 321 (1983), pp. 1273–5.
2. S. Levenstein, 'Stress and Peptic Ulcer: Life Beyond Helicobacter', *British Medical Journal*, 316 (1998), pp. 538–41.
3. *Merriam-Webster's Medical Dictionary* (Springfield, MA: Merriam-Webster Inc., 1995).
4. *Collins English Dictionary*, 6th edn (London: Harper Collins, 2003).
5. R. J. Culverwell, *Hints to the Nervous and Dyspeptic on the Causes and Cure of Nervousness, Indigestion, Haemorrhoids and Constipation* (London: for the author, 1837), p. 16.
6. As argued in A. Carden-Coyne and C. E. Forth, 'The Belly and Beyond: Body, Self and Culture in Ancient and Modern Times', in A. Carden-Coyne and C. E. Forth (eds), *Cultures of the Abdomen: Diet, Digestion and Fat in the Modern World* (New York: Palgrave Macmillan, 2005), pp. 1–11.
7. G. H. Brieger, 'Dyspepsia: The American Disease? Needs and Opportunities for Research', in C. E. Rosenberg (ed.), *Healing and History: Essays for George Rosen* (New York: Science History Publications, 1979), pp. 179–90, on pp. 188–9.
8. W. F. Bynum (ed.) *Gastroenterology in Britain: Historical Essays* (London: Wellcome Institute for the History of Medicine, 1997), p. 5.
9. R. Porter, 'Biliousness', in Bynum (ed.), *Gastroenterology in Britain*, pp. 7–28, on p. 7.
10. J. Huggett, *The Mirror of Health: Food, Diet and Medical Theory 1450–1660* (Bristol: Stuart Press, 1995); M. Schoenfeldt, 'Fables of the Belly in Early Modern England', in D. Hillman and C. Mazzio (eds), *The Body in Parts: Fantasies of Corporeality in Early Modern Europe* (New York: Routledge, 1997), pp. 243–61; K. Albaba, *Eating Right in the Renaissance* (Berkeley, CA: University of California Press, 2002); and H. M. Nunn, 'Home Bodies: Matters of Weight in Renaissance Women's Medical Manuals', in E. Klaver (ed.), *The Body in Medical Culture* (New York: Suny Press, 2009), pp. 15–36.
11. Notable exceptions to this are E. A. Williams, 'Neuroses of the Stomach: Eating, Gender, and Psychopathology in French Medicine 1800–1870', *Isis*, 98:11 (2007), pp. 54–97; and G. N. Grob, 'The Rise of Peptic Ulcer 1900–1950', *Perspectives in Biology and Medicine*, 46:4 (2003), pp. 550–66.
12. A. I. Rogers and D. Hoel, 'Peptic Ulcer Disease: Retracing Science's Journey through the Gut', *Postgraduate Medicine*, 102:5 (1997), pp. 158–67; and W. Y. Lau and C. K. Leow, 'History of the Perforated Duodenal and Gastric Ulcers', *World Journal of Surgery*, 21:8 (1997), pp. 890–6.

13. Carden-Coyne and Forth (eds), *Cultures of the Abdomen*. See also S. M. Weineck, 'Digesting the Nineteenth Century: Nietzsche and the Stomach of Modernity', *Romanticism*, 12:1 (2006), pp. 35–43.
14. J. J. Brumberg, *Fasting Girls: The Emergence of Anorexia Nervosa as a Modern Disease* (Cambridge, MA: Harvard University Press, 1988); S. Hof, *Anorexia Nervosa: The Historical and Cultural Specificity: Fallacious Theories and Tenacious 'Facts'* (Lisse: Swets & Zeitlinger, 1994); and E. Shorter, 'The First Great Increase in Anorexia Nervosa', *Journal of Social History*, 21:1 (1987), pp. 69–96. For obesity, see S. L. Gilman, *Fat: A Cultural History of Obesity* (Cambridge: Polity Press, 2008).
15. S. Shapin, 'Trusting George Cheyne: Scientific Expertise, Common Sense, and Moral Authority in Early Eighteenth-Century Dietetic Medicine', *Bulletin of the History of Medicine*, 77:2 (2003), pp. 263–97.
16. Grob, 'The Rise of Peptic Ulcer'; and D. A. Christie and E. M. Tansey (eds), *Peptic Ulcer: Rise and Fall* (London: Wellcome Trust, 2002).
17. A. Wilson, 'On the History of Disease-Concepts: The Case of Pleurisy', *History of Science*, 38:3 (2000), pp. 271–319, on p. 304.
18. R. Porter, 'Gout: Framing and Fantasizing Disease', *Bulletin of the History of Medicine*, 68:1 (1994), pp. 1–28, on pp. 3–4. See also R. Porter and G. S. Rousseau, *Gout: The Patrician's Malady* (New Haven, CT: Yale University Press, 1998). A critique of the Victorian period as being typified by high levels of deaths from epidemic diseases can be found in F. Condrau and M. Worboys, 'Second Opinions: Epidemics and Infections in Nineteenth-Century Britain', *Social History of Medicine*, 20:1 (2007), pp. 147–58.
19. J. C. Whorton, *Inner Hygiene: Constipation and the Pursuit of Health in Modern Society* (Oxford: Oxford University Press, 2000).
20. J. C. Eno, *The Stomach and its Trials* (London: F. Newbury & Sons, 1871), pp. 9–10.
21. A. Hardy, 'The Medical Response to Epidemic Disease during the Long Eighteenth Century', in J. A. I. Champion (ed.), *Epidemic Disease in London*, Centre for Metropolitan History Working Papers Series, No. 1 (London: Centre for Metropolitan History, 1993), pp. 65–70.
22. Condrau and Worboys, 'Second Opinions'.
23. Warren, 'Unidentified Curved Bacilli'.
24. Brieger, 'Dyspepsia', pp. 188–9.
25. M. T. May, *Galen on the Usefulness of the Parts of the Body* (Ithaca, NY: Cornell University Press, 1968), p. 204.
26. See C. Korsmeyer, *Making Sense of Taste: Food and Philosophy* (Ithaca, NY: Cornell University Press, 1999), pp. 13–14.
27. A. Benedetti, *Historia Corporis Humani* (1497), vol. 3, p. xiii.
28. See W. Pagel, *Joan Baptista van Helmont: Reformer of Science and Medicine* (Cambridge: Cambridge University Press, 1982), p. 95; and L. S. King, *The Philosophy of Medicine: The Early Eighteenth Century* (London: Harvard University Press, 1978), pp. 132–4.
29. M. Foucault, *The Birth of the Clinic* (Paris: Presses Universitaires de France, 1963).
30. N. D. Jewson, 'The Disappearance of the Sick-Man from Medical Cosmology', *Sociology*, 10:2 (1976), pp. 225–44.
31. C. E. Rosenberg and J. Golden (eds), *Framing Disease: Studies in Cultural History* (New Brunswick, NJ: Rutgers University Press, 1992). For a similar approach to the construction of gastric ulcer disease, see P. Thagard, *How Scientists Explain Disease* (Princeton, NJ: Princeton University Press, 1999).

32. J. H. Baron, 'Peptic Ulcer', *Mount Sinai Journal of Medicine*, 67:1 (2000), pp. 58–62; J. H. Baron and A. Sonnenberg, 'Period- and Cohort-Age Contours of Death from Gastric and Duodenal Ulcer in New York 1804–1998', *American Journal of Gastroenterology*, 43:10 (2001), pp. 2887–91; J. H. Baron, 'Hospital Admissions for Peptic Ulcer and Indigestion in London and New York in the 19th and 20th Centuries', *Gut*, 50:4 (2002), pp. 568–70; J. H. Baron, 'Publications on Peptic Ulcer in Britain, France, Germany and the US', *European Journal of Gastroenterology and Hepatology*, 214:7 (2002), pp. 711–15; J. H. Baron, 'Alimentary Diseases in the Poor and Middle Class in London 1773–1815 and in New York Poor 1797–1818', *Alimentary Pharmacology and Therapeutics*, 16:10 (2002), pp. 1709–14; J. H. Baron, F. Watson, and A. Sonnenberg, 'Three Centuries of Stomach Symptoms in Scotland', *Alimentary Pharmacology and Therapeutics*, 24:5 (2006), pp. 821–9; A. Sonnenberg, 'Causes Underlying the Birth-Cohort Phenomenon of Peptic Ulcer: Analysis of Mortality Data 1911–2000, England and Wales', *International Journal of Epidemiology*, 35:4 (2006), pp. 1090–7; and J. H. Baron and A. Sonnenberg, 'History of Dyspepsia in Scotland: Admissions to the Edinburgh Royal Infirmary 1729–1830', *Scottish Medical Journal*, 53:3 (2008), pp. 42–4.

1 The National Stomach

1. S. Whiting, *Memoirs of a Stomach* (London: W. E. Painter, 1853), p. 11; and S. Whiting, *Mémoires d'un estomac*, trans. C. H. Gros (Paris: J. B. Baillière, 1888).
2. Whiting, *Memoirs of a Stomach*, p. 81.
3. W. Harvey, *Exercitatio Anatomica de Motu Cordis et Sanguinis in Animalibus* (Francofurti: Sumptibus G. Fitzeri, 1628).
4. M. Grant, *Galen on Food and Diet* (London: Routledge, 2000).
5. G. S. Rousseau, *Nervous Acts: Essays on Literature, Culture and Sensibility* (Basingstoke: Palgrave Macmillan, 2004), pp. 35–40. See also E. Clarke and L. S. Jacyna, *Nineteenth-Century Origins of Neuroscientific Concepts* (Berkeley and Los Angeles, CA: University of California Press, 1987), pp. 212–307.
6. W. Cullen, *First Lines of the Practice of Physic* (Edinburgh: William Creech, 1777).
7. G. F. Drinka, *The Birth of Neurosis: Myth, Malady and the Victorians* (New York: Simon & Schuster, 1984), pp. 34–5.
8. See R. Whytt, *Observations on the Nature, Cause and Cure of those Disorders which have been commonly called Nervous Hypochondriac or Hysteric* (Edinburgh: J. Balfour, 1765). See also R. K. French, *Robert Whytt: The Soul and Medicine* (London: Wellcome Institute for the History of Medicine, 1969), pp. 31–45.
9. R. E. Siegel, *Galen's System of Physiology and Medicine* (Basel: Karger, 1968), pp. 360–82.
10. Clarke and Jacyna, *Nineteenth-Century Origins*, p. 313.
11. Whytt, *Observations*. Sympathy became discussed more widely within medical circles, with an illustrative example being W. P. Alison, 'Observations on the Physiological Principle of Sympathy, Chiefly in Reference to the Doctrines of Mr Charles Bell', *Transactions of the Medico-Chirurgical Society of Edinburgh*, 2 (1826), pp. 165–226. For analysis of the role of sympathy in early nineteenth-century physiological teaching, see P. Mazumdar, 'Anatomy, Physiology and Surgery: Physiology Teaching in Early Nineteenth-Century London', *Canadian Bulletin of Medical History*, 4:2 (1987), pp. 119–43.
12. J. Woodforde, *A Treatise on Dyspepsia or Indigestion*, 2nd edn (London: Longman, Hurst, Rees, Orme & Brown, 1821), pp. 53–7; and T. J. Graham, *A Treatise on Indigestion*, 3rd edn (London: Simpkin & Marshall, 1833), pp. 5–12.

13. Drinka, *The Birth of Neurosis*, p. 37.
14. T. C. Speer, *General Views Relating to the Stomach, its Fabric and Functions* (London: Longman, Hurst & Rees, 1818), p. 20; G. Budd, *On the Organic Diseases and Functional Disorders of the Stomach* (London: J. Churchill, 1855), p. 17; W. Bradshaw, *The Anatomy of Dyspepsia* (London: George Philip & Sons, 1864), p. 34; and W. Bradshaw, *Brain and Stomach or Mind and Matter* (London: W. Philip & Son, 1867), p. 45.
15. S. Perrengton, *A Popular Treatise on the Stomach with Certain New and Important Principles in the Treatment of Indigestion* (London: Charles & Edwin Layton, 1839), pp. 2–3. A critique of the idea of the stomach as an independent animal can be found in Culverwell, *Hints to the Nervous and Dyspeptic*, pp. 16–17.
16. For more on Abernethy, see J. L. Thornton, *John Abernethy: A Biography* (London: Simpkin Marshall, 1953).
17. For more on Hunter, see S. Jacyna, 'Images of John Hunter in the Nineteenth Century', *History of Science*, 21:1 (1983), pp. 85–108; F. Duchesneau, 'Vitalism in Late Eighteenth-Century Physiology: The Cases of Barthes, Blumenbach and John Hunter', in W. F. Bynum and R. Porter (eds), *William Hunter and the Eighteenth-Century Medical World* (Cambridge: Cambridge University Press, 1985), pp. 259–95; and R. Maulitz, *Morbid Appearances: the Anatomy of Pathology in the Early Nineteenth Century* (Cambridge: Cambridge University Press, 1987), pp. 114–17.
18. *The Works of John Hunter F.R.S.*, ed. J. F. Palmer, 4 vols (London: Longman, Rees, Orme, Brown, Green & Longman, 1837), vol. 1, pp. 201, 315, 338.
19. J. Abernethy, *An Enquiry into the Probability and Rationality of Mr Hunter's Theory of Life* (London: Longman, 1814).
20. J. Abernethy, *Surgical Observations on the Constitutional Origin and Treatment of Local Diseases* (Longman, Rees, Orme, Brown, Green & Longman, 1811); and 'My Book by John Abernethy', *British Journal of Surgery*, 17:67 (1930), pp. 369–72.
21. J. Abernethy, *The Abernethian Code of Health and Longevity* (London: J. Williams, 1829).
22. 'Memoirs of John Abernethy, M. R. C. S.', *Observer*, 10 October 1853, p. 7.
23. Abernethy, *The Abernethian Code*, p. 5.
24. Abernethy, *Surgical Observations*, pp. 67–8.
25. 'Parker on the Stomach in its Morbid States', *Dublin Journal of Medical Science*, 13 (1838), pp. 334–49, on p. 336.
26. 'A Cure of Consumption Induced by Indigestion, Exeter, after the Best of Routine Practice had Failed', *Trewman's Exeter Flying Post*, 20 April 1854, p. 4.
27. 'A Case of Consumption', *Trewman's Exeter Flying Post*, 23 March 1854, p. 8.
28. J. Howship, *Practical Remarks upon Indigestion* (London: Longman, Hurst, Rees, Orme, Brown, & Green, 1825), p. 59.
29. J. Johnson, *An Essay on Morbid Sensibility of the Stomach and Bowels, as the Proximate Cause of Indigestion* (London, 1827), p. 42.
30. 'Death of Sir Francis Chantrey', *Manchester Guardian*, 1 December 1841, p. 1.
31. See, for instance, C. Wightman, *A Treatise on the Sympathetic Relationship between the Stomach and the Brain* (London: Simpkin, Marshall & Co., 1840).
32. There are multiple references to this in the primary literature from this period. Particularly in-depth accounts can be found in G. Rees, *Practical Observations on Disorders of the Stomach*, 2nd edn (London: M Allen, 1811), pp. 79–96; and Culverwell, *Hints to the Nervous and Dyspeptic*, pp. 97–101.

33. W. H. Wollaston, 'The Croonian Lecture', *Philosophical Transactions of the Royal Society of London*, 100:1 (1810), pp. 1–15; and Howship, *Practical Remarks upon Indigestion*, p. 58. For more on nineteenth-century perceptions of seasickness, see J. Chapman, *Functional Diseases of the Stomach Part One: Seasickness, its Nature and Treatment* (London: Trubner & Co., 1864).
34. Johnson, *An Essay on Morbid Sensibility*, pp. 40–1; and L. Parker, *The Stomach in its Morbid States* (London: Longman, Orme, Brown, Green, 1838), p. 22.
35. See J. Reid, *Essays on Insanity, Hypochondriasis and other Nervous Affections* (London: Longman, Hurst, Rees, Orme, & Brown, 1816); Woodforde, *A Treatise on Dyspepsia*; and G. Warren, *A Discourse upon National Dietetics, as Connected with Dyspepsia, Hypochondriasis, Gout, Erysipelas, and many Diseases of this and other Christian Countries* (London, 1830).
36. 'Indigestion', *Lloyd's Weekly London Newspaper*, 27 September 1846, p. 5.
37. T. L. Brunton, *On Disorders of Digestion, their Consequences and Treatment* (London: Macmillan & Co., 1886), p. 89.
38. The theme of nationality has been much discussed and it seems that some form of British national identity had been formed by this period although regional differences occurred. See L. Colley, *Britons: Forging the Nation 1707–1838* (New Haven, CT, and London: Yale University Press, 1992).
39. For a sociological overview, see B. S. Turner, 'The Government of the Body, Medical Regimen and the Rationalisation of Diet', *British Journal of Sociology*, 33:2 (1982), pp. 254–69.
40. 'Dr Baillie's Posthumous Writings', *Medico-Chirurgical Review*, 4:8 (1826), pp. 364–79, on p. 372.
41. 'Parker on the Stomach', pp. 334–5.
42. 'Dyspepsia', *North Wales Chronicle*, 15 March 1851, p. 5.
43. *London Illustrated News*, 13 March 1886, p. 276.
44. A. E. Bridger, *The Demon of Dyspepsia* (London: Swan Sonnenschein, Lowrey & Co., 1888), p. iii.
45. 'Meditations on Dyspepsia', *Blackwood's Edinburgh Review*, 90 (1861) p. 302.
46. Bridger, *The Demon of Dyspepsia*, p. 10.
47. 'The Alimentary Function and its Disorders', *Edinburgh Medical and Surgical Journal*, 17 (1821), pp. 574–608, on p. 597.
48. 'Uwins on Indigestion', *The Times*, 27 December 1827, p. 4.
49. 'Mr Howship on Indigestion', *Medico-Chirurgical Review*, 4:7 (1826), pp. 34–43, on p. 34.
50. T. C. Allbutt, *On Visceral Neuroses* (London: J. Churchill, 1884), p. 5.
51. C. Hamlin, *Public Health and Social Justice in the Age of Chadwick* (Cambridge: Cambridge University Press, 1998), pp. 54–9.
52. For more on sensibility, see Rousseau, *Nervous Acts*, pp. 160–84.
53. Johnson, *An Essay on Morbid Sensibility*, p. 4.
54. Ibid., p. 6.
55. Ibid., p. 1.
56. A. Wear, 'Making Sense of Health and the Environment in Early Modern England', in A. Wear (ed.), *Medicine in Society* (Cambridge, Cambridge University Press, 1992), pp. 119–47, on pp. 126–9.
57. *The Times*, 30 September 1833, p. 3.
58. Hamlin, *Public Health and Social Justice*, p. 52.

59. A. P. W. Philip, *A Treatise on Indigestion and its Consequences*, 6th edn (London: Thomas & George Underwood, 1828), pp. 71–2.
60. 'Diet in Britain and France', *Manchester Guardian and British Volunteer*, 14 July 1827, p. 4.
61. C. Ó Gráda, *Ireland before and after the Famine: Explorations in Economic History 1800–1925*, 2nd edn (Manchester and New York: Manchester University Press, 1993); and C. Kinealy, *This Great Calamity: The Irish Famine 1845–1852* (Dublin: Gill & Macmillan, 1994).
62. M. Poovey, *Making a Social Body: British Cultural Formation 1830–1864* (Chicago, IL, and London: Chicago University Press, 1995), pp. 55–72.
63. *The Times*, 6 January 1846, p. 5.
64. Further references to this can be found in 'Present Condition of the Irish People', *Manchester Guardian*, 19 December 1846, p. 5. For more on Ireland and nutrition, see L. A. Clarkson and E. M. Crawford, *Feast and Famine: A History of Food and Nutrition in Ireland 1500–1920* (Oxford: Oxford University Press, 2001).
65. 'The Condition of the People of Ireland', *Freeman's Journal and Daily Commercial Advertiser*, 8 January 1846, p. 4.
66. This viewpoint is exemplified in W. Roberts, *Lectures on Dietetics and Dyspepsia* (London: Smith, Elder, 1885), p. 13.
67. Drawing upon the imagery of the body to describe relations between the individual and society was not a practice unique to this period. Phrases such as the 'body politic' had been applied since medieval times, although they became largely replaced by the term 'social body' throughout the nineteenth century. Analysis of the phrase 'body politic' can be found in D. G. Macrae, 'The Body and Social Metaphor', in J. Benthall and T. Polhemus (eds), *The Body as a Medium of Expression* (London: Allen Lane, 1975), pp. 59–73. See also J. O'Neill, *Five Bodies: The Human Shape of Modern Society* (Ithaca, NY, and London: Cornell University Press, 1985); Poovey, *Making a Social Body*, pp. 7–8; and R. Porter, *Bodies Politic: Disease, Death and Doctors in Britain 1650–1900* (London: Reaktion, 2003).
68. 'A Second Chapter on Taxation', *Manchester Guardian*, 24 April 1844, p. 6.
69. 'Address to the Trades of England', *Northern Star and National Trades' Journal*, 3 June 1848, p. 1.
70. 'The State of the Nation: A Word of Warning to the Working Classes', *Reynold's Newspaper*, 8 August 1858, p. 1.
71. 'Bread and Cheese v. Reform', *Lloyd's Weekly London Newspaper*, 12 December 1858, p. 1.
72. J. E. Clarke, *Indigestion: Its Causes and Cures* (London: James Epps & Co., 1888), p. vi.
73. See C. Lawrence, 'The Nervous System and Society in the Scottish Enlightenment', in B. Barnes and S. Shapin (eds), *Natural Order: Historical Studies of Scientific Culture* (London: Sage Publications, 1979), pp. 19–40.
74. Rousseau, *Nervous Acts*, p. 54.
75. For more on Verity, see ibid., pp. 56–7.
76. R. Verity, *Changes Produced in the Nervous System by Civilisation* (London: S. Highley, 1837). Similar arguments have been made for consumption in R. Porter, 'Consumption: Disease of the Consumer Society?', in J. Brewer and R. Porter (eds), *Consumption and the World of Goods* (London and New York: Routledge, 1993), pp. 58–81.

77. For more on stomach conditions in the early modern period, see A. Wear, *Knowledge and Practice in English Medicine: 1550–1680* (Cambridge: Cambridge University Press, 2000), pp. 169–78.
78. G. Cheyne, *The English Malady* (London, 1733).
79. Colley, *Britons*, pp. 36–7.
80. For an early critique of the effects of civilization, see C. Hall, *The Effects of Civilisation on European States* (1805; Cambridge: Chadwyck-Healey Ltd, 1995).
81. The term 'urban' is used somewhat loosely here. Boundary changes, subordinization and unregulated growth ensured that even contemporaries found it increasingly difficult to define what the nineteenth-century city or town was. See S. Gunn, *The Public Culture of the Victorian Middle Class* (Manchester and New York: Manchester University Press, 2000), pp. 11–12.
82. R. J. Morris and R. Rodger, 'An Introduction to British Urban History 1820–1914', in R. J. Morris and R. Rodger (eds), *The Victorian City: A Reader in British Urban History 1820–1914* (London and New York: Longman, 1993), pp. 1–39.
83. The literature on this is too voluminous to list here but a good starting point is E. A. Wrigley, *Continuity and Change: The Character of the Industrial Revolution in England* (Cambridge: Cambridge University Press, 1988).
84. 'Urbanity' and 'civility' are phrases typically relating to behaviour in cities. For more see A. Heller, *A Theory of Modernity* (Malden, MA, and Oxford: Blackwell Publishers, 1999), p. 154.
85. Wear, 'Making Sense of Health and the Environment', pp. 129–37.
86. Johnson, *An Essay on Morbid Sensibility*, pp. 49–50.
87. Philip, *A Treatise on Indigestion*, pp. 141–2.
88. Clarke, *Indigestion*, pp. 11–12.
89. Porter, 'Consumption', p. 69.
90. *The Times*, 12 November 1831, p. 3; and *The Times*, 1 August 1832, p. 4.
91. See 'The Experiences of M. De Lesseps', *Manchester Guardian*, 11 August 1883, p. 9.
92. See A. C. Vila, 'The Philosophe's Stomach: Hedonism, Hypochondria and the Intellectual in Enlightenment France', in Carden-Coyne and Forth (eds), *Cultures of the Abdomen*, pp. 89–104.
93. 'Suicide of Professor M'Cullagh', *Freeman's Journal and Daily Commercial Advertiser*, 26 October 1847, p. 3.
94. Bradshaw, *The Anatomy of Dyspepsia*, p. 9.
95. *The Times*, 2 February 1839, p. 6.
96. 'A Word to Machine Markers', *Manchester Times and Gazette*, 14 January 1832, p. 2.
97. 'Struggles of the Spitalfields Poor', *Era*, 13 July 1856, p. 12.
98. A. Wright, *Digestion: The Passport to Health* (London: William Rider & Son, 1883), pp. 37–8.
99. See A. Leared, *The Causes and Treatment of Imperfect Digestion* (London: John Churchill, 1860), pp. 33–5.
100. For more on luxury and ill health in the eighteenth century, see Porter, 'Consumption'.
101. For more on corsets and health, see E. Shorter, *Women's Bodies: A Social History of Women's Encounters with Health, Ill-Health and Medicine* (New Brunswick, NJ: Transaction Publishers, 1991), pp. 28–31; and P. A. Cunningham, *Reforming Women's Fashion 1850–1920* (Kent, OH: Kent State University Press, 2003).
102. 'Caution: Tight Stay Lacing', *Hull Packet and Humber Mercury*, 3 May 1831, p. 4.

103. 'Fatal Effects of Tight Lacing', *Freeman's Journal and Daily Commercial Advertiser*, 15 April 1844, p. 4.
104. 'Reminiscences of an Old Bohemian', *The Times*, 25 October 1882, p. 4.
105. For debate between the value of diet and medicine, see J. F. C. Harrison, 'Early Victorian Radicals and the Medical Fringe', in W. F. Bynum and R. Porter (eds), *Medical Fringe and Medical Orthodoxy 1750–1850* (London: Croom Helm, 1987), pp. 198–215.
106. Johnson, *An Essay on Morbid Sensibility*, pp. 69–70.
107. Culverwell, *Hints to the Nervous and Dyspeptic*, p. 6.
108. C. E. Rosenberg, *Explaining Epidemics and other Studies in the History of Medicine* (Cambridge: Cambridge University Press, 1992), pp. 24–5.
109. Whiting, *Memoirs of a Stomach*, pp. 40–77.
110. J. Dewar, *The Red Cross Series of Health Hand Books: Dyspepsia* (London: William Patterson & Co., 1891), p. 1.
111. See J. Johnson, *The Influence of Tropical Climates* (London, 1813); and Johnson, *An Essay on Morbid Sensibility*, p. 109.
112. L. Mumford, *The City in History: Its Origins, its Transformations and its Prospects* (New York: Harcourt, Brace & World, 1961).
113. See D. Spadafora, *The Idea of Progress in Eighteenth-Century Britain* (New Haven, CT, and London: Yale University Press, 1990).
114. For discussion on London as a nineteenth-century utilitarian city, see P. Hall, *Cities in Civilisation: Culture, Innovation and Urban Order* (London: Weidenfield & Nicolson, 1998), pp. 657–705.
115. For wider discussion, see B. Haley, *The Healthy Body and Victorian Culture* (Cambridge, MA, and London: Harvard University Press, 1978).
116. See J. Raban, *Soft City* (London: Fontana/Collins, 1974).
117. Porter, 'Consumption'.
118. T. Trotter, *A View of the Nervous Temperament* (London: Longman, Hurst, Rees & Orme, 1807), pp. 89–90. This refers to the perceived disgrace of the Romantic German novel. See T. Grieder, 'The German Drama in England, 1790–1800', *Restoration and Eighteenth-Century Theatre Research*, 32:2 (1964), pp. 39–50.
119. *The Times*, 22 April 1844, p. 4.
120. For fuller analysis, see Y. Akiyama, *Feeding the Nation: Nutrition and Health in Britain before World War One* (London: Tauris Academic Studies, 2008).
121. These ideas stem principally from Abernethy, *The Abernethian Code.*
122. Clarke, *Indigestion*, p. 31.
123. 'Diet', *Derby Mercury*, 30 December 1835, p. 4.
124. 'Manchester, by a Manchester Man', *Preston Guardian*, 18 June 1853, p. 3.
125. Leared, *The Causes and Treatment of Imperfect Digestion.*
126. J. C. Jackson, 'Tired Stomachs', *Dietetic Reformer and Vegetarian Messenger*, 79 (1881), p. 152.
127. Discussion of this is plentiful. Representative examples can be found in Leared, *The Causes and Treatment of Imperfect Digestion*, p. 8; and Bradshaw, *The Anatomy of Dyspepsia*, pp. 27–32.
128. Roberts, *Lectures on Dietics and Dyspepsia*, pp. 2–4.
129. Brunton, *On Disorders of Digestion*, p. 63.
130. Roberts, *Lectures on Dietics and Dyspepsia*, pp. 2–4.

131. For more on the history of salt, see P. Astrup, *Salt and Water in Culture and Medicine* (Copenhagen: Munksgaard, 1993); and M. Kurlansky, *Salt: A World History* (London: Vintage, 2003).
132. 'Condiments and Sauces', *Truth Tester, Temperance Advocate and Healthian Journal*, 1 (1847), pp. 106–7.
133. Roberts, *Lectures on Dietics and Dyspepsia*, pp. 10–11.
134. See Clarkson and Crawford, *Feast and Famine*, pp. 103–5.
135. For more on alcohol and temperance in nineteenth-century Britain, see N. Longmate, *The Waterdrinkers* (London: Hamish Hamilton, 1968); W. R. Lambert, *Drink and Sobriety in Victorian Wales 1820–1895* (Cardiff: University of Wales Press, 1983); and P. Townend, *Father Mathew, Temperance and Irish Identity* (Dublin: Irish Academic Press, 1992).
136. 'Father Mathew at Kennington and Fulham', *Observer*, 13 August 1843, p. 4.
137. 'Drink and Insanity', *Glasgow Herald*, 18 August 1866, p. 2.
138. 'A Short Sermon', *Manchester Times and Gazette*, 10 September 1842, p. 1.
139. B. Parsons and J. Marsh, *Anti-Bacchus: An Essay on the Crime, Diseases and other Evils Connected with the Use of Intoxicating Drinks* (London: John Snow, 1840), pp. 92–3.
140. Due to copyright restrictions, I have not been able to incorporate these images. However, they can be found within P. Burne, *The Teetotaller's Companion* (London: Arthur Hall & Co., 1847), pp. 2, 255, 258, 261, 264, 269, 272. This is available on the 'Making of the Modern World' database managed by Cengage.
141. Burne, *The Teetotaller's Companion*, p. 12.
142. Ibid., p. 31.

2 The Ulcerated Stomach

1. W. Brinton, *On the Pathology, Symptoms and Treatment of Ulcer of the Stomach* (London: John Churchill, 1857).
2. T. Southwell, *Medical Essays and Observations* (London: J. Knox, 1764), p. 140, who described 'a hole big enough to put ones thumb into'; and M. Baillie, *An Appendix to the First Edition of the Morbid Anatomy of Some of the Most Important Parts of the Human Body* (London: J. Johnson, 1798), pp. 54–5. Certainly, early nineteenth-century texts such as A. D. Stone, *A Practical Treatise on the Disease of the Stomach* (London: Cadell & Davies, 1806); and A. Munro, *The Morbid Anatomy of the Human Gullet, Stomach and Intestines* (Edinburgh: A. Constable & Co., 1811) make no mention of it.
3. Foucault, *The Birth of the Clinic*; and Jewson, 'The Disappearance of the Sick-Man'.
4. F. B. Alberti, *Matters of the Heart: History, Medicine and Emotion* (Oxford: Oxford University Press, 2010), pp. 61–74.
5. The key texts on British medical reform in this period are I. Waddington, *The Medical Profession in the Industrial Revolution* (Dublin: Gill & Macmillan, 1984); R. K. French and A. Wear (eds), *British Medicine in an Age of Reform* (London and New York: Routledge, 1991); I. Loudon, 'Medical Practitioners 1750–1850 and the Period of Medical Reform in Britain', in Wear (ed.), *Medicine in Society*, pp. 219–47; and I. A. Burney, 'Medicine in the Age of Reform', in A. Burns and J. Innes (eds), *Rethinking the Age of Reform: Britain 1780–1850* (Cambridge: Cambridge University Press, 2003), pp. 163–81. See also R. Porter, *Quacks: Fakers and Charlatans in English Medicine* (Stroud: Tempus, 2000).
6. 'Mr Howship on Indigestion', p. 34.

7. The obvious exception to this is W. Beaumont, *Experiments and Observations on the Gatric Juice, and the Physiology of Digestion* (Plattsburgh, NY: F. P. Allen, 1833).
8. Budd, *On the Organic Diseases*, p. 18.
9. Ibid., p. 19.
10. R. Lovell, *A Compleat History of Animals and Minerals* (Oxford: Goodwin, 1661), p. 366.
11. D. Gibbs, 'The Demon of Dyspepsia: Some Nineteenth-Century Perceptions of Disordered Digestion', in Bynum (ed.), *Gastroenterology in Britain*, pp. 29–42.
12. Johnson, *An Essay on Morbid Sensibility*, p. 1.
13. T. Mayo, *An Essay on the Influence of Temperament in Modifying Dyspepsia by Indigestion* (London, 1831), p. 2.
14. 'Dr Stokes on Chronic Gastritis and Dyspepsia', *Medico-Chirurgical Review*, 21 (1834), pp. 459–61, on p. 459.
15. The principal texts on this are E. Ackerknecht, *Medicine at the Paris Hospital 1794–1848* (Baltimore, MD: Johns Hopkins University Press, 1967); and Foucault, *The Birth of the Clinic*.
16. F. J. V. Broussais, *A Treatise on Physiology Applied to Pathology*, trans. J. Bell and R. La Roche (Philadelphia, PA: H. C. Carey & I. Lea, 1826), p. 309. For more on Broussais, see E. Ackerknecht, 'Broussais or a Forgotten Medical Revolution', *Bulletin of the History of Medicine*, 27 (1953), pp. 320–43; Ackerknecht, *Medicine at the Paris Hospital*, pp. 61–80; G. Canguilhem, *The Normal and the Pathological* (New York: Zone Books, 1989); and J. Duffin, 'Laennec and Broussais: The "Sympathetic" Duel', in C. Hannaway and A. L. Berge (eds), *Constructing Paris Medicine* (Amsterdam: Rodopi, 1998), pp. 251–74.
17. For English medical students in France, see Maulitz, *Morbid Appearances*, pp. 134–57; and T. N. Bonner, *Becoming a Physician: Medical Education in Britain, France, Germany and the United States 1750–1945* (Oxford: Oxford University Press, 1995), pp. 146–51.
18. Parker, *The Stomach in its Morbid States*.
19. Jacyna, 'Images of John Hunter'.
20. Mazumdar, 'Anatomy, Physiology and Surgery', p. 123. For Hunter's influence, see Jacyna, 'Images of John Hunter'; Maulitz, *Morbid Appearances*, pp. 114–17; and C. Lawrence, 'Medical Minds, Surgical Bodies: Corporeality and the Doctors', in C. Lawrence and S. Shapin (eds), *Science Incarnate: Historical Embodiments of Natural Knowledge* (Chicago, IL, and London: Chicago University Press, 1998), pp. 156–201.
21. J. Abernethy, *Lectures on the Theory and Practice of Surgery* (London: Longman, Rees, Orme, Brown, & Green, 1830), p. 1.
22. It is worth noting that by turning to a localized pathology, medicine was adopting a point of view already prevalent among surgeons as stated in Ackerknecht, *Medicine at the Paris Hospital*, pp. 25–6; and Bonner, *Becoming a Physician*, pp. 57–8.
23. R. Richardson, *Death, Dissection and the Destitute* (London: Routledge & Kegan Paul, 1987).
24. See Howship, *Practical Remarks upon Indigestion*.
25. Maulitz, *Morbid Appearances*, pp. 162–3. For more on Johnson, see A. Desmond, *The Politics of Evolution: Morphology, Medicine and Reform in Radical London* (Chicago, IL, and London: Chicago University Press, 1989), pp. 193–8.
26. For more on the decline of the traditional patient–physician relationship, see M. E. Fissell, 'The Disappearance of the Patient's Narrative and the Invention of Hospital Medicine', in French and Wear (eds), *British Medicine in an Age of Reform*, pp. 92–109.

27. J. Hope, *A Treatise on the Diseases of the Heart and Great Vessels* (London: John Churchill, 1832).
28. W. E. E. Conwell, *A Treatise on the Functional and Structural Changes of the Liver* (London: James Duncan, 1835).
29. M. A. V. G. Bolvin and A. Dugès, *A Practical Treatise on the Diseases of the Uterus and its Appendages* (London: Sherwood, Gilbert & Piper, 1834).
30. For more on Abercrombie, see H. N. Moyer, 'John Abercrombie: The First Neurologist', *Bulletin of the Society of the Medical History of Chicago*, 2:3 (1920), pp. 137–45; J. R. Watson, 'John Abercrombie and the Diagnosis of Duodenal Ulcer', *Annals of Medical History*, 3:4 (1942), pp. 468–72; and J. Pitman, 'The John Abercrombie Collection', *Proceedings of the Royal College of Physicians of Edinburgh*, 21:3 (1991), pp. 349–54.
31. J. Abercrombie, *Pathological and Practical Researches on Diseases of the Stomach*, 2nd edn (Edinburgh: Waugh & Innes, 1830), pp. iv–v.
32. Ibid., p. 11.
33. Ibid., p. 17.
34. Ibid., p. 47.
35. 'Dr Abercrombie on the Diseases of the Stomach', *Edinburgh Medical and Surgical Journal*, 31 (1829), pp. 135–60, on p. 160.
36. 'Diseases of the Stomach', *Medico-Chirurgical Review*, 10 (1829), pp. 328–54, on p. 328.
37. Occasional references are made to stomach worms causing holes in the abdominal wall. See 'Perforation of the Stomach Probably by a Worm', *Lancet*, 27 (1836), pp. 368–70.
38. Desmond, *The Politics of Evolution*, p. 9. See also M. Brown, 'Medicine, Reform and the "End" of Charity in Early Nineteenth-Century England', *English Historical Review*, 125:511 (2010), pp. 1353–88.
39. A. Burns, 'Observations on the Digestion of the Stomach after Death', *Edinburgh Medical and Surgical Journal*, 6 (1810), pp. 129–38.
40. F. L. Holmes, 'The Physical Sciences in the Life Sciences', in M. J. Nye (ed.), *The Modern Physical and Mathematical Sciences* (Cambridge: Cambridge University Press, 2002), pp. 219–36, on pp. 224–6.
41. See R. Réaumur, 'Sur la Digestion des Oiseaux', *Memoires de l'Academie Royale des Sciences*, 1756 (1752), pp. 266–307, 461–95; and L. Spallanzani, *Dissertations Relative to the Natural History of Animals and Vegetables* (London: John Murray, 1784), pp. 375–91.
42. E. Stevens, *Dissertatio Physiologica Inauguralis de Alimentarium Concoctione* (Edinburgh: Balfour & Smellie, 1777).
43. See L. Gmelin and F. Tiedemann, *Die Verdauung nach Versuchen* (Heidelberg: Karl Gros, 1827); and Beaumont, *Experiments and Observations*. See also F. L. Holmes, *Claude Bernard and Animal Chemistry: The Emergence of a Scientist* (Cambridge, MA: Harvard University Press, 1974), pp. 141–214; and S. Normandin, 'Claude Bernard and an Introduction to the Study of Experimental Medicine: Physical Vitalism, Dialetic and Epistemology', *Journal of the History of Medicine and Allied Sciences*, 62:4 (2007), pp. 495–528.
44. See W. Prout, *Chemistry, Meteorology and the Function of Digestion Considered with Reference to Natural Theology* (London: W. Pickering, 1834).
45. J. Hunter, 'On the Digestion of the Stomach after Death', *Philosophical Transactions of the Royal Society of London*, 62 (1772), pp. 447–54.
46. Budd, *On the Organic Diseases*, pp. 21–2.

47. R. Carswell, 'An Inquiry on the Chemical Solution or Digestion of the Coats of the Stomach after Death', *Edinburgh Medical and Surgical Journal*, 34 (1830), pp. 282–311, on p. 283.
48. For more on the new role of the medical professional in the courtroom, see R. Smith, *Trial by Medicine: Insanity and Responsibility in Victorian Trials* (Edinburgh: Edinburgh University Press, 1981); and I. A. Burney, *Bodies of Evidence: Medicine and the Politics of the English Inquest 1830–1926* (Baltimore, MD: Johns Hopkins University Press, 1999).
49. For toxicology, see I. A. Burney, *Poison, Detection and the Victorian Imagination* (Manchester: Manchester University Press, 2006).
50. A. S. Taylor, 'On Perforations of the Stomach from Poisoning and Disease', *Guy's Hospital Reports*, 5 (1839), pp. 8–62. See also A. S. Taylor, *A Manual of Medical Jurisprudence* (London: John Churchill, 1844), pp. 57–62.
51. Taylor, 'On Perforations of the Stomach', pp. 29–35.
52. T. Williamson, 'An Attempt to Estimate some of the Characteristic Marks by which to Judge of the Cause of Perforations of the Stomach II', *Dublin Journal of Medical Science*, 19 (1841), pp. 191–219, on p. 202.
53. E. Young, *On Perforating Ulcer of the Stomach* (London: Simpkin, Marshall, & Co., 1849).
54. 'Post-Mortem Examination: A Drunkard's Body', *Temperance Lancet and Journal of Useful Intelligence*, 5:1 (1841), pp. 35–6.
55. *National Temperance Magazine*, 1 (1844), pp. 508–9.
56. 'The Death-Bed of the Inebriate', *People's Abstinence Standard and True Social Reformer*, 1 (1849), p. 228.
57. Parsons and Marsh, *Anti-Bacchus*, p. 28.
58. 'Temperance Festival in Blackburn', *Manchester Temperance Reporter and Journal of Progress*, 18:1 (1850), pp. 142–3.
59. 'Excessive Brandy Drinking: Perforation of the Stomach after Death', *Medical Times and Gazette*, 19 (1849), p. 242.
60. 'Mr Howship on Indigestion', p. 34.
61. As late as 1850, reports can still be found of stomach ulcers being aggravated by the excessive use of purgative medicine. See 'Chronic Ulcer of the Stomach', *Philosophical Transactions of the Royal Society of London*, 140 (1850), pp. 211–12.
62. 'Treatment of Dyspepsia and Gastrodynia', *Dublin Journal of Medical and Chemical Science*, 2 (1832), pp. 143–4, on p. 144.
63. E. Ackerknecht, *Therapeutics: From the Primitives to the Twentieth Century* (London: Macmillan, 1973), p. 115.
64. 'Gastrology', *Medico-Chirurgical Review*, 9 (1828), pp. 217–18.
65. 'Dr Abercrombie on the Pathology of the Stomach, the Pancreas, and the Spleen', *Edinburgh Medical and Surgical Journal*, 21 (1824), pp. 1–15, on p. 13.
66. H. M. Hughes, 'Case of Supposed Spontaneous Perforation of the Stomach', *Guy's Hospital Reports*, 4 (1846), pp. 332–42.
67. E. Ray, 'Case of Ulcer in the Stomach Leading to Perforation and Death in Nineteen Hours', *Guy's Hospital Reports*, 4 (1846), pp. 343–50.
68. 'Chronic Ulcer of the Stomach', *American Journal of Medical Science*, 20 (1850), pp. 35–6.
69. Although the author is rarely written about today, Budd was proposing theories regarding food deficiency diseases as far back as 1842. See R. E. Hughes, 'George Budd

(1808–1882) and Nutritional Deficiency Diseases', *Medical History*, 17:2 (1973), pp. 127–35, on p. 127.

70. Budd, *On the Organic Diseases*, pp. 108–21.
71. See above, p. 37.
72. Budd, *On the Organic Diseases*, pp. 69–71.
73. Brinton, *On the Pathology, Symptoms and Treatment of Ulcer of the Stomach*, p. 188.
74. Ibid., p. 120.
75. 'Bayard on Diseases of the Stomach', *Dublin Quarterly Journal of Medical Science*, 34 (1862), pp. 342–9, on pp. 342–3.

3 The Laboratory Stomach

1. An earlier draft of this chapter was published in *Journal of the History of Medicine and Allied Sciences* in July 2009.
2. Réaumur, 'Sur la Digestion de Oiseaux'.
3. Spallanzani, *Dissertations Relative to the Natural History of Animals and Vegetables*, pp. 375–91.
4. W. Coleman and F. L. Holmes (eds), *The Investigative Enterprise: Experimental Physiology in Nineteenth-Century Medicine* (Berkeley, CA: University of California Press, 1988), pp. 4–5. See also S. Jacyna, 'The Laboratory and the Clinic: The Impact of Pathology on Surgical Diagnosis in the Glasgow Western Infirmary 1875–1910', *Bulletin of the History of Medicine*, 62:3 (1988), pp. 384–406.
5. B. Moynihan, *Duodenal Ulcer* (Philadelphia, PA, and London: W. B. Saunders Co., 1910).
6. For a general overview of the relationship between science and medicine in this period, see W. F. Bynum, *Science and the Practice of Medicine in the Nineteenth Century* (Cambridge: Cambridge University Press, 1994).
7. G. L. Geison, *Michael Foster and the Cambridge School of Physiology: The Scientific Enterprise in Late Victorian Society* (Princeton, NJ: Princeton University Press, 1978); and R. G. Frank, *Harvey and the Oxford Physiologists: Scientific Ideas and Social Interaction* (Berkeley, CA, and London: University of California Press, 1980).
8. J. E. Lesch, *Science and Medicine in France: The Emergence of Experimental Physiology 1790–1855* (Baltimore, MD: Johns Hopkins University Press, 1967).
9. C. Lawrence, 'Incommunicable Knowledge: Science, Technology and the Clinical Art in Britain', *Journal of Contemporary History*, 20:4 (1985), pp. 503–20.
10. The British antivivisection movement has been well documented in R. D. French, *Antivivisection and Medical Science in Victorian Society* (London and Princeton, NJ: Princeton University Press, 1976); C. Lansbury, *The Old Brown Dog: Women, Workers and Vivisection in Edwardian England* (Madison, WI: University of Wisconsin Press, 1985); S. Richards, 'Drawing the Life-Blood of Physiology: Vivisection and the Physiologists' Dilemma, 1870–1900', *Annals of Science*, 43:11 (1986), pp. 27–56; N. A. Rupke (ed.), *Vivisection in Historical Perspective* (London: Croom Helm, 1987); S. Hamilton (ed.), *Animal Welfare and Anti-Vivisection 1870–1910: Nineteenth Century Women's Mission* (New York: Routledge, 2004); and P. White, 'Sympathy Under the Knife: Experimentation and Emotion in Late Victorian Medicine', in F. B. Alberti (ed.), *Medicine, Emotion and Disease 1700–1950* (Basingstoke: Palgrave Macmillan, 2006), pp. 100–24.
11. W. H. Brock, *Justus von Liebig: The Chemical Gatekeeper* (Cambridge: Cambridge University Press, 1997); J. Morrell, *The Chemist Breeders: The Research Schools of Liebig and*

Thomas Thomson (Cambridge: Heffer, 1972); and M. R. Finlay, 'Quackery and Cookery: Justus von Liebig's Extract of Meat and the Theory of Nutrition in the Victorian Age', *Bulletin of the History of Medicine*, 66:3 (1992), pp. 404–18.

12. C. Bernard, *An Introduction to the Study of Experimental Medicine*, trans. H. C. Greene (New York: Macmillan, 1927).
13. Lesch, *Science and Medicine in France*, pp. xi–xii.
14. J. H. Warner, 'The Fall and Rise of Professionalism: Epistemology, Authority and the Emergence of Laboratory Medicine in Nineteenth-Century America', in A. Cunningham and P. Williams (eds), *The Laboratory Revolution in Medicine* (Cambridge: Cambridge University Press, 1992), pp. 110–41, on p. 125.
15. Roberts, *Lectures on Dietetics and Dyspepsia*.
16. G. O. Drewry, *Common-Sense Management of the Stomach* (London: Henry S. King & Co., 1875).
17. Clarke, *Indigestion*.
18. It is worth noting that when urging scientific programmes into clinical practice, physiologists were more likely to stress the simplicity of such methods than their complexities, as argued in J. H. Warner, 'Ideals of Science and their Discontents in Late Nineteenth-Century American Medicine', *Isis*, 82:3 (1991), pp. 454–78, on p. 458.
19. The best overview of this can be found in A. L. Gillespie, *A Manual of Modern Gastric Methods: Chemical, Physical and Therapeutical* (Edinburgh: Oliver & Boyd, 1899).
20. M. Einhorn, *Diseases of the Stomach* (London: Baillière & Co., 1897), p. 80.
21. W. G. Morgan, 'Some Experiences with the Einhorn Duodenal Bucket and a Modified Thread Test', *American Journal of Medical Science*, 141:5 (1911), pp. 649–58.
22. Gillespie, *A Manual of Modern Gastric Methods*, p. 111. See also Einhorn, *Diseases of the Stomach*.
23. G. Herschell, 'A New Gastro-Diaphane', *Lancet*, 163 (1904), p. 1361.
24. Gillespie, *A Manual of Modern Gastric Methods*, p. 120; and G. Herschell, 'An Improved Apparatus for Auto-Lavage of the Stomach', *Lancet*, 164 (1904), p. 532.
25. See F. Riegel, *Diseases of the Stomach* (Philadelphia, PA, and London: W. B. Saunders, 1903), p. 38; and F. Riegel, 'Diagnosis of Gastric Disease', *Lancet*, 147 (1896), p. 568.
26. Gillespie, *A Manual of Modern Gastric Methods*, pp. 124–31.
27. Leared, *The Causes and Treatment of Imperfect Digestion*, p. 174.
28. T. H. Bast, *The Life and Times of Adolf Kussmaul* (New York: P. B. Hoeber, 1926), p. 100.
29. C. A. Ewald, *Diseases of the Stomach* (Edinburgh: Young J. Pentland, 1892), p. 600.
30. For a detailed description of these, see Gillespie, *A Manual of Modern Gastric Methods*, pp. 11–12.
31. Lesch, *Science and Medicine in France*, p. 10.
32. Lawrence, 'Incommunicable Knowledge'.
33. Richards, 'Drawing the Life Blood of Physiology'.
34. See, for instance, 'Scientific Medicine', *Society for Protection of Animals from Vivisection Pamphlets 2* (London: Office of the Society for Protection of Animals from Vivisection, 1881–2), pp. 1–2; and 'The Scientist at the Bedside', *Zoophilist*, 12 (1882), pp. 1–4.
35. Warner, 'Ideals of Science'.
36. S. Richards, 'Vicarious Suffering, Necessary Pain: Physiological Method in Late Nineteenth-Century Britain', in Rupke (ed.), *Vivisection in Historical Perspective*, pp. 125–48.
37. White, 'Sympathy under the Knife', p. 117.
38. Warner, 'Ideals of Science', p. 457.
39. French, *Antivivisection and Medical Science*.

40. For more on Starling, see J. Henderson, *A Life of Ernest Starling* (New York and Oxford: Oxford University Press, 2005).
41. E. H. Starling, *Recent Advances in the Physiology of Digestion* (London: Archibald Constable & Co., 1906), p. 71.
42. Ibid., pp. 63–4. Starling adopted these techniques from the Russian physiologist Ivan Pavlov. See D. P. Todes, *Pavlov's Physiology Factory: Experiment, Interpretation, Laboratory Enterprise* (Baltimore, MD: Johns Hopkins University Press, 2002).
43. Starling, *Recent Advances*, pp. 65–71, 72.
44. 'Notes and Notices', *Animals Defender and Zoophilist*, 35 (1915), p. 53.
45. L. L. Hageby and L. K. Schartau, *The Shambles of Science* (London: Ernest Bell, 1903), pp. 25–6.
46. C. Bolton, *Ulcer of the Stomach* (London: Edward Arnold, 1913), p. v.
47. E. Berdge, 'Torture of Animals in London Today', *Animals' Defender and Zoophilist*, 35 (1915), p. 15. The original article can be found in C. Bolton, 'Recent Observations on the Pathology of Ulcer of the Stomach with Indications for Treatment', *British Medical Journal*, 1 (1915), pp. 707–10.
48. 'Samples of Vivisection', *Society for Protection of Animals from Vivisection Pamphlets* (London: Office of the Society for Protection of Animals from Vivisection, 1896), p. 2.
49. 'Some 1896 Vivisections', *Society for Protection of Animals from Vivisection Pamphlets* (London: Office of the Society for Protection of Animals from Vivisection, 1897), pp. 1–2; and 'Samples of Vivisection', *Society for Protection of Animals from Vivisection Pamphlets* (London: Office of the Society for Protection of Animals from Vivisection, 1895), p. 1.
50. M. J. Durey, 'Bodysnatchers and Benthamites: The Implications of the Dead Body Bill for the London Schools of Anatomy 1820–42', *London Journal*, 2 (1976), pp. 200–25; and Richardson, *Death, Dissection and the Destitute*.
51. See, for instance, 'The Scientific Use of Hospitals', *Nineteenth Century*, 49 (1901), pp. 57–63. See also S. E. Lederer, *Subjected to Science: Human Experimentation in America before the Second World War* (Baltimore, MD, and London: Johns Hopkins University Press, 1995); and Burney, *Bodies of Evidence*, pp. 147–9.
52. Robert G. Frank has used the example of methods of graphic representation of the heart to illustrate this in R. G. Frank Jr, 'The Telltale Heart: Physiological Instruments, Graphic Methods and Clinical Hopes 1854–1914', in Coleman and Holmes (eds), *The Investigative Enterprise*, pp. 211–90.
53. 'The Acids of the Stomach', *British Medical Journal*, 2 (1889), pp. 774–5.
54. W. M. Ord, 'On the Diagnosis and Treatment of Gastric Ulcer', *Dublin Journal of Medical Science*, 88 (1889), pp. 545–62.
55. Gillespie, *A Manual of Modern Gastric Methods*, pp. 47–8.
56. H. S. Souttar and T. Thompson, 'The Direct Inspection of the Gastric Mucous Membrane', *Quarterly Journal of Medicine*, 1:4 (1908), pp. 376–9.
57. W. Hill, *On Gastroscopy and Oesophago-Gastroscopy* (London: John Bale, 1912), p. 2. In fact, death as a result of the misuse of the tube appears to have been rare although some cases were detailed in *Cleveland Journal of Medicine*, 4 (1899), p. 445; and *Journal of the American Medical Association*, 24 (1895), p. 911.
58. 'Contribution to the Determination of the Percussion Limits of the Stomach', *Dublin Journal of Medical Science*, 3 (1887), pp. 21–40.
59. Riegel, *Diseases of the Stomach*, pp. 52–3.
60. Leared, *The Causes and Treatment of Imperfect Digestion*, p. 221.

61. G. A. Herschell, *Manual of Intragastric Technique* (London: H. J. Glaisher, 1903), pp. 3–13.
62. Ibid., pp. 26–7.
63. Ibid., pp. 3–13.
64. G. M. Niles, *The Diagnosis and Treatment of Digestive Diseases* (London: Henry Kimpton, 1914), p. 170.
65. Herschell, *Manual of Intragastric Technique*, p. 3.
66. E. A. Williams, 'Gags, Funnels and Tubes: Forced Feeding of the Insane and of Suffragettes', *Endeavour*, 32:4 (2008), pp. 134–40.
67. For more on the early use of the tube, see J. Friedenwald, 'Note on the Discovery and Early Use of the Stomach Tube', *Medical Life*, 34:12 (1927), pp. 639–44; J. Friedenwald, 'The History of the Development of the Stomach Tube with Some Notes on the Duodenal Tube', *Bulletin of the History of Medicine*, 4:6 (1936), pp. 425–54; and R. H. Major, 'History of the Stomach Tube', *Annals of Medical History*, 6 (1934), pp. 500–9.
68. D. A. Moxey, 'Feeding by the Nose in Attempted Suicide by Starvation', *Lancet*, 100 (1872), pp. 444–6.
69. D. A. Moxey, 'Feeding by the Nose in Attempted Suicide by Starvation cont...', *Lancet*, 100 (1872), pp. 489–90.
70. Brumberg, *Fasting Girls*, pp. 104–5.
71. For fuller discussion of the relationship between the medical profession and forcible feeding, see J. F. Geddes, 'Culpable Complicity: The Medical Profession and the Forcible Feeding of Suffragettes, 1909–1914', *Women's History Review*, 17:1 (2008), pp. 79–94.
72. Accounts of forcible feedings in prisons were reported internationally. See D. Barnes, 'How it Feels to be Forcibly Fed', *New York World Magazine*, 6 September 1914, pp. 5, 6, 17.
73. For more on Kitty Marion, see J. Holledge, *Innocent Flowers: Women in the Edwardian Theatre* (London: Virago Press, 1981), pp. 56–7.
74. 'A House Defaced and a Window Broken', *The Times*, 5 November 1909, p. 4.
75. 'Woman Suffrage: The Fabian Society and Forcible Feeding', *The Times*, 19 November 1909, p. 10.
76. 'Woman Suffrage', *The Times*, 12 November 1909, p. 12.
77. M. A. Elston, 'Women and Anti-Vivisection in Victorian England, 1870–1900', in Rupke (ed.), *Vivisection in Historical Perspective*, pp. 259–94.
78. 'Against Forcible Feeding', *Suffragette*, 2 (1913), p. 194.
79. 'The Imprisonment of Lady Constance Lytton', *The Times*, 26 January 1910, p. 10.
80. 'Home Office Statements Refuted', *Votes for Women*, 5 (1912), p. 664.
81. F. W. Pethick Lawrence, 'Newcastle Prisoners Released: The Story of their Ordeal in Prison', *Votes for Women*, 3 (1909), p. 67.
82. For more on the role of prison doctors, see J. Sim, *Medical Power in Prisons: Prison Medical Service in England 1774–1988* (Buckingham: Open University Press, 1990); and A. Hardy, Development of the Prison Medical Service 1774–1895', in R. Creese, W. F. Bynum and J. Bearn (eds), *The Health of Prisoners: Historical Essays* (Amsterdam: Rodopi, 1995), pp. 59–82.
83. 'The Imprisonment of Lady Constance Lytton'; and 'The Law and the Women', *Common Cause*, 7 (1916), p. 570. A fuller account can be found in C. Lytton, *Prisons and Prisoners: Some Personal Experiences* (London: Heinemann, 1914).
84. 'Pried Open Teeth of Suffragettes' *New York Times*, 17 October 1909, p. C4.
85. 'Suffragette who was Forcibly Fed 232 Times', *Manchester Guardian*, 17 April 1914, p. 9.

86. 'The Case of William Ball', *Manchester Guardian*, 10 May 1912, p. 4.
87. 'The Attempt at Suicide', *The Times*, 29 June 1912, p. 6.
88. J. Purvis, 'The Prison Experiences of the Suffragettes in Edwardian Britain', *Women's History Review*, 4:1 (1995), pp. 103–33.
89. J. R. Walkowitz, *Prostitution and Victorian Society: Women, Class and the State* (New York and Cambridge: Cambridge University Press, 1982), pp. 56–7, 201–2.
90. Lansbury, *The Old Brown Dog*, p. 99.
91. 'Miss Lilian Lenton's Story of her Experience', *Manchester Guardian*, 22 October 1913, p. 10.
92. 'Forcible Feeding: Opinions of Medical Experts, Grave Danger to Life Involved', *Votes for Women*, 3:82 (1909), p. 3.
93. 'The Woman Suffragists', *The Times*, 29 September 1909, p. 10.
94. 'A Physician on Forcible Feeding', *The Times*, 26 February 1913, p. 10.
95. 'Fasting Prisoners and Compulsory Feeding', *British Medical Journal*, 2 (1909), pp. 997–8.
96. C. Mansell-Moullin, 'To the Editor of the Times', *The Times*, 29 September 1909, p. 10. It is worth noting that he had a somewhat sympathetic attitude towards the campaign, having operated upon Emily Davison after she was knocked unconscious at the Derby racetrack in 1913 and being an active member of the Men's League for Women's Suffrage. See 'Edith Mansell-Moullin', in E. Crawford, *The Women's Suffrage Movement: A Reference Guide 1866–1928* (New York: Routledge, 2001), p. 375.
97. 'The Feeding of Suffragist Prisoners', *The Times*, 7 October 1909, p. 5.
98. V. Horsley, 'Forcible Feeding', *The Times*, 21 December 1909, p. 10. For more on Horsley's opposition to force-feeding, see Geddes, 'Culpable Complicity'.
99. It is worth noting that the use of the tube continued to cause controversy throughout the remainder of the decade due to its use on Sinn Féin prisoners, the most noteworthy episode in this being the death of Thomas Ashe. See T. Ashe, *The Death of Thomas Ashe* (Dublin: J. M. Butler, 1917).
100. 'Forcible Feeding of Suffragists' *Manchester Guardian*, 29 September 1909, p. 4.

4 The Surgical Stomach

1. R. S. Stevenson, *Famous Illnesses in History* (London: Eyre & Spottiswoode, 1962), pp. 32–43.
2. See F. L. Donaldson, *King George VI and Queen Elizabeth* (London: Weidenfield & Nicolson), p. 15; and P. Howarth, *George VI: A New Biography* (London: Hutchinson, 1987), p. 253.
3. H. Ballatine, 'Remarks on Duodenal Ulcer with Notes of a Case', *Edinburgh Medical Journal*, 14 (1903), pp. 532–5, on p. 532.
4. A comparative case study has been provided by Dale C. Smith who argued that in American practice few practitioners saw appendicitis as an exclusively surgical disease around 1890, yet by 1920 the surgical approach was perceived as dominant both professionally and publicly. Surgeons utilized this success to explain and campaign for the wider validity and acceptance of general surgical judgement and intervention, as well as the validity, professionalism and safety of their field. See D. C. Smith, 'Appendicitis, Appendectomy and the Surgeon', *Bulletin of the History of Medicine*, 70:3 (1996), pp. 414–41.
5. See A. Dally, *Fantasy Surgery 1880–1930 with Special Reference to Sir William Arbuthnot Lane* (Amsterdam: Rodopi, 1996); H. Ellis, 'Some British Contributions to Abdominal

Surgery 1870–1910', in Bynum (ed.), *Gastroenterology in Britain*, pp. 29–43; and Whorton, *Inner Hygiene*, pp. 55–79.

6. Lawrence, 'Incommunicable Knowledge'.
7. O. H. Wangensteen and S. D. Wangensteen, *The Rise of Surgery from Empiric Craft to Scientific Discipline* (Folkestone: Dawson, 1978); and G. H. Brieger, 'From Conservative to Radical Surgery in Late Nineteenth-Century America', in C. Lawrence (ed.), *Medical Theory, Surgical Practice: Studies in the History of Surgery* (Routledge: London and New York, 1992), pp. 216–31.
8. J. G. Smith, *Abdominal Surgery* (London: John Churchill, 1891).
9. B. Moynihan, *The Pathology of the Living and Other Essays* (Philadelphia, PA, and London: W. B. Saunders Co., 1910).
10. Abernethy, *Surgical Observations*; and Abernethy, *The Abernethian Code*.
11. See Mazumdar, 'Anatomy, Physiology and Surgery'.
12. See J. Hunter, *A Treatise on the Blood, Inflammation and Gun-Shot Wounds* (London: John Richardson, 1794). It is notable that no abdominal complaints are featured in the data provided within S. T. Anning, 'The Practice of Surgery in Leeds 1823–1824', *Medical History*, 23:1 (1979), pp. 59–95. For more on early nineteenth-century surgery, see P. Stanley, *For Fear of Pain: British Surgery 1790–1850* (Amsterdam and New York: Rodopi, 2003).
13. *The Times*, 30 January 1824, p. 2; and *The Times*, 3 February 1824, p. 3. For more on Cooper, see Desmond, *The Politics of Evolution*, pp. 111–14.
14. A. Cooper and B. Travers, *Surgical Essays: Part One* (London: Cox, 1818), pp. 111–45. See also Stanley, *For Fear of Pain*, pp. 76–7.
15. See W. T. G. Morton, *Remarks on the Proper Mode of Administering Sulphuric Ether by Inhalation* (Boston, MA, 1847). For historical analysis, see M. Pernick, *A Calculus of Suffering: Pain, Professionalism and Anaesthesia in Nineteenth-Century America* (New York: Columbia University Press, 1985); R. J. Wolfe, *Tarnished Idol: William Thomas Green Morton and the Introduction of Surgical Anaesthesia: A Chronicle of the Ether Controversy* (San Anselmo, CA: Norman Publishing, 2001); and T. Dormandy, *The Fight Against Pain* (New Haven, CT: Yale University Press, 2006).
16. R. Druitt, *The Principles and Practice of Modern Surgery*, 3rd edn (Philadelphia, PA: Blanchard & Lea, 1852), pp. 420–8.
17. *Monthly Journal of Medical Science*, 10 (1848), p. 186; and J. C. Cooper, 'Description of the Operation of Gastrotomy', *Guy's Hospital Reports*, 4 (1858), pp. 13–18.
18. L. Tait, *Diseases of Women and Abdominal Surgery* (Leicester: Richardson & Co., 1889), p. v.
19. See G. Southam, *Ovariotomy: Removal of an Encysted Tumour of the Left Uterine Appendages* (Salford: W. F. Jackson, 1845); J. Y. Simpson, *Ovariotomy: Is it or is it not an Operation Justifiable upon the Common Principles of Surgery?* (Edinburgh: Sutherland & Knox, 1846); and R. Lee, 'Analysis and Summary of One Hundred and Sixty Two Cases of Ovariotomy or Excision of the Ovarium which have Taken Place in Great Britain', *Edinburgh Medical and Surgical Journal*, 78 (1852), pp. 70–95.
20. For primary literature on hysterectomy, see J. K. Thornton, *Cases of Hysterectomy: With Remarks on the Value of the Carbolic Acid Spray in this Operation* (London: John Bale & Sons, 1885); K. Thomas, *Contributions to the Surgical Treatment of Tumours of the Abdomen* (Edinburgh: Oliver & Boyd, 1885); W. J. Sinclair, 'Vaginal Hysterectomy for Cancer', *Practitioner*, 43:6 (1889), pp. 413–34; and J. Bland-Sutton, *Essays on Hysterectomy* (London: Adlard, 1904).

21. A. Digby, 'Women's Biological Straitjacket', in S. Mendus and J. Rendall (eds), *Sexuality and Subordination: Interdisciplinary Studies of Gender in the Nineteenth Century* (London and New York: Routledge, 1989), pp. 192–220.
22. A. Dally, *Women under the Knife: A History of Surgery* (London: Hutchison Radius, 1991).
23. Stanley, *For Fear of Pain*, p. 78.
24. N. Chevers, 'An Inquiry into Certain of the Causes of Death after Injuries and Surgical Operations in London Hospitals', *Guy's Hospital Reports*, 2 (1843), pp. 78–101, on p. 94.
25. J. D. Malcolm, *The Physiology of Death from Traumatic Fever: A Study in Abdominal Surgery* (London: J. & A. Churchill, 1893).
26. A fuller account of criticism and wariness towards anaesthesia and discussion of the perceived benefits of pain can be found in Pernick, *A Calculus of Suffering*, pp. 35–76, while its benefits are discussed on pp. 77–92. For a recent account of the use of anaesthesia in nineteenth-century Britain, see S. J. Snow, *Operations without Pain: the Practice and Science of Anaesthesia in Victorian Britain* (Basingstoke: Palgrave Macmillan, 2006).
27. J. E. Erichsen, *The Science and Art of Surgery being a Treatise on Surgical Injuries, Diseases and Operations*, 5th edn (London: James Walton, 1869), p. 16.
28. Ibid., pp. 443–59.
29. H. A. Hare and E. Q. Thornton, 'A Study of the Influence of Chloroform upon the Respiration and Circulation', *Lancet*, 142 (1893), pp. 996–9. See also T. Spencer Wells, *Modern Abdominal Surgery* (London: J. & A. Churchill, 1891), pp. 7–8.
30. Druitt, *The Principles and Practice of Modern Surgery*, p. 428.
31. C. H. Talamon, *Appendicitis and Perityphlitis* (Edinburgh and London: Young J. Pentland, 1893), pp. 1–2.
32. For nephrectomy, see G. Elder, *Three Cases of Nephrectomy with Remarks on the Operation* (London: John Bale & Sons, 1885); and C. Williams, *Carcinoma of the Kidney, Nephrectomy: A Case of Gastrostomy for Malignant Disease of the Oesophagus* (London, 1892).
33. D. Newman, *Lectures to Practitioners on the Diseases of the Kidney Amenable to Surgical Treatment* (London and New York: Longmans, Green & Co., 1888).
34. T. Billroth, *Die Allgemeine Chirurgische Pathologie and Chirurgie* (Berlin: G. Reimer, 1863). See also L. M. Zimmerman and I. Veith, *Great Ideas in the History of Surgery* (1961; Novato, CA: Norman Publishing, 1993), pp. 188–98.
35. H. Engel, 'Billroth, Christian Albert Theodor', in C. C. Gillispie (ed.), *Dictionary of Scientific Biography 2* (New York: Charles Scribner's Sons, 1970), pp. 129–31.
36. F. F. Cartwright, *The Development of Modern Surgery* (London: Arthur Baker Ltd, 1967), pp. 196–7.
37. N. C. Dobson, 'Abdominal Section in Perforating Ulcer of the Stomach', *Bristol Medico-Chirurgical Review*, 7 (1883), p. 196.
38. Notable early examples include 'Perforation of Gastric Ulcer and its Treatment by Abdominal Section and Suture', *British Medical Journal*, 1 (1892), pp. 63–4; H. Gilford, 'A Case of Perforated Gastric Ulcer for which Gastrorrhaphy was Performed: Death on the Thirty-First Day', *British Medical Journal*, 1 (1893), pp. 944–6; G. Barling, 'The Treatment of Perforated Gastric Ulcer with Report of Successful Drainage in a Case', *British Medical Journal*, 1 (1893), pp. 1258–9; 'Perigastric Abscess Perforating the Diaphragm and Causing Empyema; Operation; Death; Necropsy', *Lancet*, 142 (1893), pp. 89–91; 'Operation in Perforation of Gastric Ulcer', *Practitioner*, 52 (1894), pp. 452–3; R. H. Bourchier Nicholson, 'Case of Perforative Ulcer of Stomach Treated by Lapa-

rotomy, Stitching and Washing Out', *British Medical Journal*, 2 (1894), pp. 982–3; and A. Marmaduke Shield, 'Two Cases of Ulcer of the Duodenum in which Laparotomy was Performed, with Remarks on Ulcers of the Duodenum', *Lancet*, 145 (1895), pp. 1169–74.

39. A. R. Parsons, 'The Desirability of Operative Interference in Suspected Perforation of Chronic Ulcer of the Stomach', *Dublin Journal of Medical Science*, 94 (1892), pp. 26–35, on p. 32.
40. Ibid., p. 26.
41. H. Gilford, 'The Surgical Treatment of (Unperforated) Gastric Ulcer with an Account of Three Cases in which Operations were Performed', *Guy's Hospital Reports*, 53 (1898), pp. 103–25.
42. S. Trombley, *Sir Frederick Treves: The Extra-ordinary Edwardian* (Routledge: London, 1989), p. 86; and C. Lawrence, 'Democratic, Divine and Heroic: The History and Historiography of Surgery', in Lawrence (ed.), *Medical Theory, Surgical Practice*, pp. 1–47, on pp. 1–2.
43. 'Recent Advances in the Surgery of the Stomach', *Lancet*, 167 (1906), pp. 609–10.
44. See D. Bateman, *Berkeley Moynihan: Surgeon* (London: Macmillan & Co., 1940).
45. A. W. Mayo-Robson and B. Moynihan, *Diseases of the Stomach and their Surgical Treatment*, 2nd edn (London: Ballière, Tindall & Cox, 1904), p. v.
46. See M. Worboys, *Spreading Germs: Diseases, Theories and Medical Practice in Britain 1865–1900* (Cambridge and New York: Cambridge University Press, 2000).
47. B. Moynihan, *The Advance of Medicine* (Oxford: Clarendon Press, 1932), pp. 23–5.
48. See U. Tröhler, 'To Operate or Not to Operate: Scientific and Extraneous Factors in Therapeutical Controversies within the Swiss Society of Surgery 1913–1988', in W. F. Bynum and V. Nutton (eds), *Essays in the History of Therapeutics* (Amsterdam: Rodopi, 1991), pp. 89–114.
49. Wilde and Hurst have also noted that many early twentieth-century surgeons persistently generated new surgical knowledge and worked within a culture that valued innovation. See S. Wilde and G. Hurst, 'Learning from Mistakes: Early Twentieth-Century Surgical Practice', *Journal of the History of Medicine and Allied Sciences*, 64:1 (2009), pp. 38–77.
50. See O. Temkin, 'The Scientific Approach to Disease: Specific Entity and Individual Sickness', in A. C. Crombie (ed.), *Scientific Change: Historical Studies in the Intellectual, Social and Technical Conditions for Scientific Discovery and Technical Innovation from Antiquity to the Present* (New York: Basic Books, 1963), pp. 629–47.
51. A. Carless, 'The Surgery of the Stomach', *Practitioner*, 12 (1900), pp. 491–501. See also J. M. Purser, 'On the Modern Diagnosis of Diseases of the Stomach', *Dublin Journal of Medical Science*, 90:6 (1890), pp. 449–87. Most notably, work undertaken into diabetes and the pancreas had encouraged this process. See V. Harley, 'The Normal Absorption of Fat and the Effect of Extirpation of the Pancreas on It', *Journal of Physiology*, 18:1 (1895), pp. 1–14; V. Harley, *Pancreatic Diabetes in Animals and Man* (Manchester, 1895); and A. W. Mayo-Robson and B. Moynihan, *Diseases of the Pancreas and their Surgical Treatment* (Philadelphia, PA, and London: W. B. Saunders & Co., 1902).
52. For more on Pavlov, see Todes, *Pavlov's Physiology Factory*.
53. 'The Effects of Extirpation of the Pancreas', *Lancet*, 136 (1890), p. 1407.
54. The idea that the intestines were also an evolutionary remnant no longer needed by modern man, and which could be surgically removed, has been analysed in Dally, *Fantasy Surgery*, pp. 109–22; and Whorton, *Inner Hygiene*, pp. 55–79.

55. C. Schlatter, 'Oesophagy-Enterostomy after Total Extirpation of the Stomach', *Lancet*, 151 (1898), pp. 141–6; and C. Schlatter, 'Further Observations on a Case of Total Extirpation of the Stomach in the Human Subject', *Lancet*, 152 (1898), pp. 1314–18. See also B. Moynihan, 'On Total Extirpation of the Stomach with a Record of an Unsuccessful Case', *British Medical Journal*, 2 (1903), pp. 1458–9. A fuller list of attempted extirpations of the stomach can be found in Mayo-Robson and Moynihan, *Diseases of the Stomach*, pp. 123–8.
56. See C. Lawrence and G. Weisz (eds), *Greater than the Parts: Holism in Biomedicine, 1920–1950* (Oxford: Oxford University Press, 1998).
57. 'Recent Advances in the Surgery of the Stomach', p. 610.
58. See Ackerknecht, *Medicine at the Paris Hospital*, pp. 25–6; Mazumdar, 'Anatomy, Physiology and Surgery', p. 119; and Bonner, *Becoming a Physician*, pp. 57–8. For more on early nineteenth-century pathology, see Maulitz, *Morbid Appearances*. The classic paper on increasing medical specificity is Jewson, 'The Disappearance of the Sick-Man'.
59. B. Moynihan, *Two Lectures on Gastric and Duodenal Ulcer* (Bristol: John Wright & Sons, 1923), p. 47.
60. Ibid., p. 219.
61. Ibid., p. 124.
62. Ibid., p. 46.
63. Moynihan, *The Pathology of the Living*, p. 2.
64. Ibid., p. 2.
65. Mayo-Robson and Moynihan, *Diseases of the Stomach*, pp. 7–9.
66. Ibid., p. 9.
67. Jacyna, 'The Laboratory and the Clinic', pp. 390–2.
68. Abernethy, *Lectures on the Theory and Practice of Surgery*, p. 1.
69. A. E. Maylard, *A Treatise on the Surgery of the Alimentary Canal: Comprising the Oesophagus, the Stomach, the Small and Large Intestines, and the Rectum* (London: J. & A. Churchill, 1896), p. 174.
70. A. W. Mayo-Robson, 'Abstracts of the Hunterian Lectures on the Surgery of the Stomach', *British Medical Journal*, 1 (1900), pp. 562–6, on pp. 562–3.
71. A. E. Barker, 'On Some Cases of Operation for Chronic Non-Malignant Disease of the Stomach', *Lancet*, 160 (1902), pp. 501–3; 'The Surgery of Chronic Dyspepsia', *Lancet*, 160 (1902), pp. 512–13; T. K. Dalziel, 'Thirty Cases of Gastro-Enterostomy for Non-Malignant Affections of the Stomach', *Lancet*, 160 (1902), pp. 503–4; and G. Barling, 'An Address on the Diagnosis and Surgical Treatment of Certain Cases of Chronic Indigestion', *British Medical Journal*, 2 (1904), pp. 1623–7.
72. A. B. Mitchell, 'Contribution to a Discussion on the Treatment of Gastric Ulcer from a Surgical Standpoint', *Dublin Journal of Medical Science*, 123 (1907), pp. 179–80. For more on exploratory surgery, see A. E. Maylard, 'An Address on the Surgery of the Stomach with Illustrative Cases', *Lancet*, 156 (1900), pp. 1253–8, on p. 1254.
73. R. Saundby, 'Remarks on the Indications for Operative Interference in Diseases of the Stomach', *British Medical Journal*, 2 (1904), pp. 1621–3, on p. 1621.
74. B. Dawson, 'The Diagnosis of Gastric Ulcer', *British Medical Journal*, 2 (1905), pp. 1032–4.
75. P. R. Griffiths, 'Perforating Gastric Ulcer with Notes of Two Successful Cases after Operation', *British Medical Journal*, 1 (1900), pp. 572–4, on pp. 572–3.
76. Mayo-Robson and Moynihan, *Diseases of the Stomach*, p. 14.

77. F. B. Jessett, *Surgical Diseases and Injuries of the Stomach and Intestines* (London: Baillière & Co., 1892), p. 6.
78. Ibid., p. 9.
79. Marmaduke Shield, 'Two Cases of Ulcer of the Duodenum'.
80. 'Reviews and Bibliographical Notices', *Dublin Journal of Medical Science*, 131 (1910), pp. 132–3, on p. 132.
81. Maylard, *A Treatise on the Surgery of the Alimentary Canal*, p. 277.
82. See, for instance, C. E. Reeves, *Diseases of the Stomach and Duodenum* (London: Simpkin, Marshall, 1856), pp. 277–354.
83. S. Fenwick, *The Morbid States of the Stomach and Duodenum and their Relation to the Diseases of other Organs* (London: John Churchill & Sons, 1868), pp. 5, 64, 132.
84. Moynihan, *Duodenal Ulcer*; B. Moynihan, 'Duodenal Ulcer', *Practitioner*, 78 (1907), pp. 747–61, on p. 747.
85. R. H. Fitz, 'Acute Pancreatitis', *Boston Medical and Surgical Journal*, 120:8 (1889), pp. 181–7; and Mayo-Robson and Moynihan, *Diseases of the Stomach*, pp. 58–60.
86. Mayo-Robson and Moynihan, *Diseases of the Stomach*, p. 45.
87. See A. W. Mayo-Robson, 'Abstracts of the Hunterian Lectures on the Surgery of the Stomach II', *British Medical Journal*, 1 (1900), pp. 624–9, on p. 626; and Mayo-Robson and Moynihan, *Diseases of the Stomach*, pp. 338–77. The idea that humans could be born with two stomachs had often caused controversy and local excitement in the nineteenth century. See, 'The Boy with Two Stomachs', *British Medical Journal*, 1 (1877), p. 669.
88. Moynihan, *Duodenal Ulcer*, pp. 129–30.
89. Ibid., p. 11.
90. Discussion of this can be found in W. C. Alexander, 'A Case Illustrating the Difficulty of Diagnosis in Gastric Ulcer', *British Medical Journal*, 1 (1897), p. 1345; and G. S. Middleton, 'Some Observations on the Difficulties of Diagnosis in Abdominal Disease, with Brief Accounts of Illustrative Cases', *Glasgow Medical Journal*, 5 (1897), pp. 322–39.
91. A. B. Mitchell, 'A Contribution to the Surgery of Perforated Gastric Ulcer', *British Medical Journal*, 1 (1900), pp. 567–75, on p. 569.
92. B. Moynihan, 'A Hunterian Lecture on Some Problems of Gastric and Duodenal Ulcer', *British Medical Journal*, 1 (1923), pp. 221–6, on p. 221.
93. 'Book Reviews', *Annals of Surgery*, 52:4 (1910), pp. 572–5, on pp. 572–3.
94. A. W. Mayo-Robson, 'The Hunterian Lecture on Duodenal Ulcer and its Treatment', *British Medical Journal*, 1 (1907), pp. 248–54, on p. 249.
95. Moynihan, *Duodenal Ulcer*, p. 11.
96. Ibid., p. 101.
97. Ibid., pp. 101–28.
98. See, for instance, C. Perry and L. E. Shaw, 'Malignant Disease of the Stomach', *Guy's Hospital Reports*, 58 (1904), pp. 121–52, on pp. 125–6.
99. Moynihan, *The Pathology of the Living*, p. 47.
100. Moynihan, *Duodenal Ulcer*, pp. 101–2.
101. Moynihan, *The Pathology of the Living*, p. 50.
102. Ibid., p. 7.
103. I. W. Wheeler. 'The Diagnosis and Treatment of Gastric and Duodenal Ulcers', *Irish Journal of Medical Science*, 1 (1923), pp. 1–12, on p. 12.
104. H. Maclean, *Modern Views on Digestion and Gastric Disease* (London: Constable & Co. Ltd, 1925), p. v.
105. T. C. Hunt, *The Common Causes of Chronic Indigestion: Differential Diagnosis and Treatment* (London: Balliere, Tindall & Cox, 1933), pp. 1–2.

106. 'The Treatment of Gastric Ulcer', *British Medical Journal*, 1 (1908), pp. 387–91, on p. 390.
107. 'The Therapeutic Value of Hydrochloric Acid in Diseases of the Stomach', *British Medical Journal*, 2 (1905), pp. 891–7, on pp. 893–4.
108. B. W. Sippy, 'Gastric and Duodenal Ulcer: Medical Cure by an Efficient Removal of Gastric Juice Corrosion', *Journal of the American Medical Association*, 64 (1915), pp. 1625–30.
109. H. Maclean, 'A Lecture on the Intensive Alkaline Treatment of Gastric and Duodenal Ulcer', *British Medical Journal*, 1 (1928), pp. 619–23, on p. 619.
110. J. L. Steven, 'On the Surgical Treatment of Diseases of the Stomach from a Physician's Point of View', *Lancet*, 163 (1904), pp. 1487–93, on p. 1487.
111. A. R. Parsons, 'The Diagnosis and Treatment of a Perforated Gastric Ulcer, with Notes on a Successful Case', *Dublin Journal of Medical Science*, 121 (1906), pp. 81–92, on pp. 86–7.
112. R. Saundby, 'A Clinical Lecture on Ulcer of the Stomach', *British Medical Journal*, 1 (1900), pp. 121–2, on p. 122.
113. G. Herschell, *The Surgical Treatment of Duodenal Ulcer* (London: Henry J. Glashier, 1910), pp. 38–9.
114. M. J. Smyth, 'Gastro-Jejunal Ulcer', *Irish Journal of Medical Science*, 9 (1931), p. 552.
115. W. H. C. Romanis, 'The Surgical Aspect of Gastric Ulcer', *Practitioner*, 55 (1923), p. 232.
116. A. Hurst, 'New Views on the Pathology, Diagnosis and Treatment of Gastric and Duodenal Ulcer', *British Medical Journal*, 1 (1920), pp. 559–63.
117. B. Moynihan, 'On the Recognition of some Acute Abdominal Diseases', *Practitioner*, 63 (1931), p. 7.
118. Moynihan, 'A Hunterian Lecture', p. 223.
119. R. Mailer, 'The Later Results of Operations on the Stomach and Duodenum', *Glasgow Medical Journal*, 108 (1936), p. 110.
120. O. E. I. Sampson, 'Chronic Duodenal Ulcer and Perforation', *British Medical Journal*, 2 (1944), pp. 864–9, on p. 866.
121. Moynihan, *The Pathology of the Living*, p. 355.
122. A. E. Barclay, *The Stomach and Oesophagus: A Radiographic Study* (London: Sherratt & Hughes, 1913).
123. See F. Taylor, 'Examination of Stomach by X-Rays and Tube Filled with Subnitrate of Bismuth', *British Medical Journal*, 1 (1905), p. 720; A. C. Jordan, 'Duodenal Obstruction as Shown by Radiography', *British Medical Journal*, 1 (1911), pp. 1172–4; and G. A. Pirie, 'The Diagnosis of Disease of the Stomach by the X-Rays', *Edinburgh Medical Journal*, 9 (1912), pp. 137–42, on p. 137.
124. W. C. Bosanquet and H. S. Clogg, *The Stomach, Intestines and Pancreas* (London: John Bale, Sons & Danielsson Ltd, 1909), p. iii.
125. Saundby, 'Remarks on the Indications for Operative Interference', pp. 1621–2.
126. R. Saundby, *The Treatment of Diseases of the Digestive System* (London: Charles Griffin & Co., 1906), p. 1.
127. Ibid., pp. 74–5.
128. 'A Discussion on the Surgical Treatment of the Non-Malignant Diseases of the Stomach', *British Medical Journal*, 2 (1905), pp. 767–810, on p. 772.
129. A. F. Hurst, 'Duodenal Ulcer and the Hypertonic Stomach', *British Medical Journal*, 1 (1920), p. 688.

5 The Psychosomatic Stomach

1. An earlier draft of this chapter was published in *Medical History* in January 2010.
2. Abernethy, *Surgical Observations*.
3. Analysis of the medical complaints of this conflict has been somewhat limited so far as academic attention has frequently been directed to conflicts such as the First World War. See M. Harrison, *Medicine and Victory: British Military Medicine in the Second World War* (Oxford: Oxford University Press, 2004), p. 1.
4. Lawrence and Weisz (eds), *Greater than the Parts*, pp. 1–2. For more on early twentieth-century holistic medicine, see S. W. Tracy, 'George Draper and American Constitutional Medicine 1916–1946: Re-Inventing the Sick Man', *Bulletin of the History of Medicine*, 66:1 (1992), pp. 53–89; and C. Timmermann, 'Constitutional Medicine, Neoromanticism, and the Politics of Antimechanism in Interwar Germany', *Bulletin of the History of Medicine*, 75:1 (2001), pp. 27–39; and D. Cantor (ed.), *Reinventing Hippocrates* (Aldershot: Ashgate, 2001).
5. This argument was expressed in Niles, *The Diagnosis and Treatment of Digestive Diseases*, pp. 282, 377. A similar viewpoint can be found slightly earlier in W. V. Valzah and J. D. Nisbet, *The Diseases of the Stomach* (London: Rebman Publishing Co., 1899), pp. 7–8.
6. W. S. Fenwick, *Dyspepsia: Its Varieties and Treatment* (Philadelphia, PA, and London: W. B. Saunders, 1910), pp. 18, 19.
7. Ibid., pp. 18, 20.
8. Lawrence and Weisz, *Greater than the Parts*, pp. 1–4.
9. See W. B. Cannon, 'The Influence of Emotional States on the Functions of the Alimentary Canal', *American Journal of Medical Science*, 137:4 (1909), pp. 480–6; W. B. Cannon, *The Mechanical Factors of Digestion* (London: Edward Arnold, 1911); W. B. Cannon, *Bodily Changes in Pain, Hunger, Fear and Rage: An Account of Recent Researches* (New York and London: D. Appleton & Co., 1915); W. B. Cannon, *Traumatic Shock* (New York: D. Appleton & Co., 1923); and W. B. Cannon, *The Wisdom of the Body* (New York: Norton, 1932).
10. W. C. Alvarez, *Nervous Indigestion* (London: William Heinemann, 1930).
11. H. Cushing, 'Peptic Ulcer and the Interbrain', *Surgery, Gynaecology and Obstetrics*, 55:1 (1932), pp. 1–34.
12. F. Alexander, 'The Influence of Psychologic Factors upon Gastrointestinal Disturbances', *Psychoanalytic Quarterly*, 3 (1934), pp. 501–88. For more, see F. Alexander, *Psychosomatic Medicine* (New York and London: Norton, 1950).
13. Lawrence and Weisz (eds), *Greater than the Parts*, pp. 1–4. Holism, however, was not entirely new. For more on the relationship between nineteenth- and twentieth-century thinking on disease, see E. Shorter, *From Paralysis to Fatigue: A History of Modern Psychosomatic Medicine* (New York: Free Press, 1992); E. Shorter, *From the Mind into the Body: The Cultural Origins of Psychosomatic Symptoms* (New York: Free Press, 1994); and D. Cantor, 'The Diseased Body', in R. Cooter and J. V. Pickstone (eds), *Medicine in the Twentieth Century* (Amsterdam: Harwood Academic, 2000), pp. 347–64.
14. It is worth noting that 'anxiety' was employed as a somewhat vague expression which in many ways had replaced neurasthenia as an umbrella term for a variety of mental disorders. See G. W. G. James, 'Anxiety Neurosis', *Lancet*, 236 (1940), pp. 561–4.
15. D. T. Davies and A. T. Macbeth Wilson, 'Observations on the Life-History of Chronic Peptic Ulcer', *Lancet*, 230 (1937), pp. 1353–60, on p. 1353.

16. For contemporary statistical evidence of ulcer as a male disease, see F. D. Jennings, 'Perforated Peptic Ulcer: Changes in Age-Incidence and Sex-Distribution in the Last 150 Years', *Lancet*, 235 (1940), pp. 395–8, 444–7.
17. Alexander, 'The Influence of Psychologic Factors'.
18. Davies and Macbeth Wilson, 'Observations on the Life-History of Chronic Peptic Ulcer', p. 1360.
19. For more on social medicine, see D. Porter (ed.), *Social Medicine and Medical Sociology in the Twentieth Century* (Amsterdam: Rodopi, 1997).
20. J. Busfield, *Managing Madness: Changing Ideas and Practices* (London: Ulman Hyman, 1986), pp. 326–57.
21. For shell shock, see A. Young, *The Harmony of Illusions: Inventing Post-Traumatic Stress Disorder* (Princeton, NJ: Princeton University Press, 1995), pp. 50–85; A. Babington, *Shell-Shock: A History of the Changing Attitudes to War Neurosis* (Barnsley: Cooper, 1997); B. Shepherd, *A War of Nerves: Soldiers and Psychiatrists 1914–1994* (Cambridge, MA: Harvard University Press, 2001); and P. Leese, *Shell Shock: Traumatic Neurosis and the British Soldiers of the First World War* (New York: Palgrave, 2002). For Gulf War syndrome, see J. Wheelwright, *The Irritable Heart: The Medical Mystery of the Gulf War* (New York and London: Norton, 2001); and M. Pall, *Explaining 'Unexplained Illness'* (New York: Binghamton, 2007).
22. Modern military combat regularly encourages situations that involve possibly the most intense forms of stress known to human beings, not least because of the constant imminent, unpredictable threat to soldier's lives. Groups of medically unexplained symptoms arising during times of conflict have been given the generic names of 'war syndromes' or 'post-combat disorders'. See K. C. Hyams, F. S. Wignall and R. Roswell, 'War Syndromes and their Evaluation: From the US Civil War to the Persian Gulf War', *Annals of Internal Medicine*, 125:5 (1996), pp. 398–405; W. J. Coker, B. M. Bhatt, N. J. Blatchley and J. T. Graham, 'The Clinical Findings of the first 1000 Gulf War Veterans in the Ministry of Defence's Medical Assessment Programme', *British Medical Journal*, 318 (1999), pp. 290–4; and E. Jones and S. Wessely, 'War Syndromes: The Impact of Culture on Medically Unexplained Symptoms', *Medical History*, 49:1 (2005), pp. 55–78.
23. A. Hurst, *Medical Diseases of War*, 4th edn (London: Edward Arnold & Co., 1944), p. 194.
24. See J. D. Howell, 'Soldier's Heart: The Redefinition of Heart Disease and Speciality Formation in Early Twentieth-Century Great Britain', *Medical History*, Supplement 5:1 (1985), pp. 1–33; C. Lawrence, 'Moderns and Ancients: The "New Cardiology" in Britain 1880–1930', *Medical History*, Supplement 5:1 (1985), pp. 1–33; and C. F. Wooley, *The Irritable Heart of Soldiers and the Origins of Anglo-American Cardiology* (Aldershot: Ashgate, 2002).
25. A wider discussion of how this was an international problem is available in T. L. Cleve, *The Saccharine Disease* (Bristol: John Wright, 1974). Detailed statistics of the British experience can be found in J. N. Morris and R. M. Titmuss, 'Epidemiology of Peptic Ulcer Vital Statistics', *Lancet*, 244 (1944), pp. 841–5.
26. B. Bager, 'Beitral Zur Kenntnis über Vorkommen, Klinink und Behandlung von Perforierten Magenund Duodenalgeschuwüren Nebst Einer Untersuchung über die Spatresultate Nach Verschiedenen Operationsmethoden', *Acta Chirurgica Scandinavica*, 64 (1929), Supplement 11.
27. Jennings, 'Perforated Peptic Ulcer'.

28. G. G. Taylor, 'The Abdominal Surgery of Total War', *Glasgow Medical Journal*, 19:6 (1942), pp. 123–42, on pp. 133–4.
29. R. T. Payne and C. Newman, 'Interim Report on Dyspepsia in the Army', *British Medical Journal*, 2 (1940), pp. 619–21.
30. P. H. Wilcox, 'Gastric Disorders in the Services', *British Medical Journal*, 1 (1940), pp. 1008–12.
31. J. G. Graham and J. D. O. Kerr, 'Digestive Disorders in the Force', *British Medical Journal*, 1 (1941), pp. 473–6.
32. C. A. Hinds-Howell, 'A Review of Dyspepsia in the Army', *British Medical Journal*, 2 (1941), pp. 473–4.
33. Payne and Newman, 'Interim Report on Dyspepsia'.
34. This was also noted later on in H. Tidy, 'Peptic Ulcer and Dyspepsia in the Army', *British Medical Journal*, 2 (1943), pp. 473–7, on p. 473.
35. For more on the strong stomach, willpower and military manhood, see A. Carden-Coyne, 'American Guts and Military Manhood', in Carden-Coyne and Forth (eds), *Cultures of the Abdomen*, pp. 71–86. See also O. A. Trowell, 'The Relation of Tobacco Smoking to the Incidence of Chronic Duodenal Ulcer', *Lancet*, 223 (1934), pp. 808–9; and V. S. Hodson, 'Duodenal Ulcers and Cigarette Smoking', *Lancet*, 228 (1936), pp. 1235–6.
36. Wilcox, 'Gastric Disorders in the Services'.
37. G. Pegge, 'Psychiatric Casualties in First Days of War', *British Medical Journal*, 2 (1939), pp. 764–5.
38. This forms part of a wider debate regarding the impact of panic upon the general population. See R. M. Titmuss, *Problems of Social Policy* (London: His Majesty's Stationary Office; Longmans, Green & Co., 1950); R. Mackay, *Half the Battle: Civilian Morale in Britain during the Second World War* (Manchester: Manchester University Press, 2002).
39. 'Psychiatric Casualties in London, September 1940', *British Medical Journal*, 2 (1940), pp. 553–5.
40. C. C. Spicer, D. N. Stewart and D. M. de R. Winser, 'Incidence of Perforated Peptic Ulcer: Effect of Heavy Air Raids', *Lancet*, 239 (1942), pp. 259–61; and C. C. Spicer, D. N. Stewart and D. M. de R. Winser, 'Perforated Peptic Ulcer during the Period of Heavy Air-Raids', *Lancet*, 243 (1944), p. 14.
41. Spicer et al., 'Perforated Peptic Ulcer'. It is worth noting that similar research conducted into the Glasgow air-raids seemed to contradict these findings. See C. F. W. Illingworth and L. D. W. Scott, 'Acute Perforated Peptic Ulcer: Frequency and Incidence in the West of Scotland', *British Medical Journal*, 2 (1944), pp. 617–18.
42. Spicer et al., 'Incidence of Perforated Peptic Ulcer'.
43. H. Seyle, 'Perforated Peptic Ulcer during Air-Raid', *Lancet*, 241 (1943), p. 252.
44. This viewpoint was expressed in 'Total War and the Individual', *Lancet*, 237 (1941), pp. 791–2. For a contemporary text which dealt with these themes, see M. E. Rehfuss, *Indigestion: Its Diagnosis and Management* (Philadelphia, PA, and London: W. B. Saunders Co., 1943), p. 346.
45. Payne and Newman, 'Interim Report on Dyspepsia'.
46. An illustrative example of investigators pronouncing these points can be found in Illingworth and Scott, 'Acute Perforated Peptic Ulcer'.
47. R. Coope, 'Recent Trends in Gastro-Enterological Treatment', *Practitioner*, 149:5 (1942), pp. 277–83, on p. 277.
48. R. A. M. Scott, 'Incidence of Peptic Ulcer', *British Medical Journal*, 1 (1945), p. 457.

49. Tidy, 'Peptic Ulcer and Dyspepsia', p. 473. The reference to drivers probably stems from apparent links made between gastric illness and the medical problems commonly suffered by London bus drivers. See A. B. Hill, 'An Investigation into the Sickness Experience of London Transport Workers, with Special Reference to Digestive Disturbances', *Medical Research Council International Health Research Board Report* (London: His Majesty's Stationary Office, 1937).
50. A. H. Douthwaite, 'Gastro-Enterology', *Practitioner*, 147 (1941), pp. 622–9, on p. 624.
51. Morris and Titmuss, 'Epidemiology of Peptic Ulcer Vital Statistics'.
52. 'Recent Advances in Treatment', *Practitioner*, 151 (1943), pp. 258–62, on pp. 259–60.
53. Harrison, *Medicine and Victory*.
54. See D. French, *Raising Churchill's Army: The British Army and the War against Germany 1919–1945* (Oxford: Oxford University Press, 2000), pp. 242–6.
55. Harrison, *Medicine and Victory*, pp. 1–3.
56. J. B. W. Rowe, 'Wartime Diet for Peptic Ulcer Patients', *British Medical Journal*, 2 (1943), p. 464.
57. J. J. Horwich, 'Incidence of Peptic Ulcer', *British Medical Journal*, 2 (1944), p. 866.
58. Rehfuss, *Indigestion*, pp. 359–61.
59. F. Twigg, 'Perforation of Duodenal Ulcer While Flying', *British Medical Journal*, 2 (1930), p. 687.
60. Rehfuss, *Indigestion*, pp. 356–61.
61. Hurst's key contribution to gastric ulcer was A. F. Hurst and M. J. Stewart, *Gastric and Duodenal Ulcer* (London and New York: H. Milford, Oxford University Press, 1929).
62. Hurst, *Medical Diseases of War*, pp. 176–8.
63. W. Brockbank, 'The Dyspeptic Soldier: A Record of 931 Consecutive Cases', *Lancet*, 239 (1942), pp. 39–42, on p. 41.
64. Tidy, 'Peptic Ulcer and Dyspepsia', p. 473.
65. It is also interesting to note that many of those advocating 'ulcer battalions' were particularly interested in the potential for research into the gastric system that could take place in such units. See F. R. Brown, 'Duodenal Ulcer Battalions', *British Medical Journal*, 2 (1942), p. 530.
66. See www.bbc.co.uk/ww2peopleswar/stories/06/a8790906.shtml [accessed 21 April 2009].
67. Psychologists, for instance, created instruments intended to screen out men most likely to succumb to battle exhaustion, keeping so-called 'psychoneurotics' out of combat roles. See J. T. Copp and B. McAndrew, *Battle Exhaustion: Soldiers and Psychiatrists in the Canadian Army 1939–1945* (Montreal: McGill-Queen's University Press, 1990); and N. Thalassis, 'The Use of Intelligence Testing in the Recruitment of "Other Ranks" in the Armed Forces During the Second World War', *History of Philosophy and Psychology*, 5 (2003), pp. 17–29.
68. A. Hurst, 'Wartime Diet for Peptic Ulcer Patients', *British Medical Journal*, 2 (1943), p. 523.
69. For comparison with appendicitis, see Smith, 'Appendicitis, Appendectomy and the Surgeon'.
70. Suggestions that surgery should be widely employed to allow a quick return to service was criticized in I. Bennett, 'Peptic Ulcer in the Services', *British Medical Journal*, 1 (1940), p. 458.
71. H. Tidy, 'Peptic Ulcer', *Practitioner*, 153:4 (1944), pp. 197–203, on p. 201.
72. Bennett, 'Peptic Ulcer in the Services'.

73. Coope, 'Recent Trends in Gastro-Enterological Treatment', p. 277.
74. B. Gilsenan, 'Dyspepsia of Peptic Ulcer Type and its Relationship to Personality Type and Anxiety', *Practitioner*, 157:2 (1946), pp. 457–8.
75. A. Lewis, 'The Psychological Aspects of Indigestion', *Practitioner*, 153:4 (1944), pp. 257–60, on p. 260.
76. Hurst, *Medical Diseases of War*, pp. 176–8, 193.
77. Ibid., pp. 176–8, 193–5.

Concluding Remarks

1. The history of stress is an area of enquiry which has only recently begun to develop at the time of the writing, although an overview can be found in C. L. Cooper, *Stress: A Brief History* (Oxford: Blackwell, 2004).
2. An official history of this can be found in Smith Kline and French International Co., *The H2 Receptor Anthology: The Discovery of Histamine H2-Receptors and their Antagonists* (Welwyn Garden City: Smith Kline & French, 1982).
3. For a fuller discussion, see Christie and Tansey (eds), *Peptic Ulcer*, pp. 95–112.
4. Baron, 'Peptic Ulcer'; Baron and Sonnenberg, 'Period- and Cohort-Age Contours of Death'; Baron, 'Hospital Admissions for Peptic Ulcer'; Baron, 'Publications on Peptic Ulcer'; Baron, 'Alimentary Diseases in the Poor and Middle Class'; Baron et al., 'Three Centuries of Stomach Symptoms in Scotland'; Sonnenberg, 'Causes Underlying the Birth-Cohort Phenomenon of Peptic Ulcer'; and Baron and Sonnenberg, 'History of Dyspepsia in Scotland'.
5. C. S. Goodwin, 'Duodenal Ulcer, Campylobacter Pylori and the Leaking Roof Concept', *Lancet*, 332 (1988), pp. 467–9.
6. L. Zanten and V. Zanten, 'The Aging Stomach or the Stomachs of the Ages', *Gut*, 41:4 (1997), pp. 575–6.
7. Baron and Sonnenberg, 'Period- and Cohort-Age Contours of Death'.
8. See M. Susser, 'Period Effects, Generation Effects and Age Effects in Peptic Ulcer Mortality', *Journal of Chronic Disease*, 35:1 (1982), pp. 29–40, on p. 38.
9. Baron, 'Hospital Admissions for Peptic Ulcer and Indigestion'.
10. See F. B. Alberti, 'Angina Pectoris and the Arnolds: Emotions and Heart Disease in the Nineteenth Century', *Medical History*, 52:2 (2008), pp. 221–36.
11. See E. S. Watkins, 'Medicine, Masculinity and the Disappearance of Male Menopause in the 1950s', *Social History of Medicine*, 21:2 (2008), pp. 329–44.
12. See Wilson, 'On the History of Disease-Concepts', p. 304. For further discussion of retrospective diagnosis, see J. Arrizabalaga, 'Medical Causes of Death in Preindustrial Europe: Some Historiographical Considerations', *Journal of the History of Medicine and Allied Sciences*, 54:2 (1999), pp. 241–60, on pp. 256–7; J. Arrizabalaga, 'Problematising Retrospective Diagnosis in the History of Disease', *Asclepio*, 54:1 (2002), fasc. 1, pp. 51–70; and A. W. Bates, *Emblematic Monsters: Unnatural Conceptions and Deformed Births in Early Modern Europe* (Amsterdam and New York: Rodopi, 2005), pp. 175–97.
13. See Rosenberg and Golden (eds), *Framing Disease*, pp. xiii–xxvi.
14. Wooley, *The Irritable Heart of Soldiers*, p. 68.
15. Baron, 'Publications on Peptic Ulcer'.
16. W. F. Bynum and J. C. Wilson, 'Periodical Knowledge: Medical Journals and their Editors in Nineteenth-Century Britain', in W. F. Bynum, S. Lock, and R. Porter (eds),

Medical Journals and Medical Knowledge: Historical Essays (London: Routledge, 1992), pp. 9–48.

17. The first description being B. Travers, 'Additional Observations on a Report of Rupture of the Stomach and Escape into the Cavity of the Abdomen by Crampton', *Medico-Chirurgical Transactions*, 8:1 (1817), pp. 228–45, on p. 228.
18. The trend for counting has been continued in more recent epidemiological research, with the number of MD theses in Edinburgh being unproblematically correlated with dyspepsia admissions in Baron et al., 'Three Centuries of Stomach Symptoms in Scotland'.
19. Thagard, *How Scientists Explain Disease*, p. i.
20. Ibid., pp. 114–17.
21. G. D. Yeats, *Some Observations on the Duodenum or Second Stomach* (London: G. Woodfall, 1820), pp. 1–13.
22. Abercrombie, *Pathological and Practical Researches*, pp. 103–7.
23. See, for instance, 'Ulceration of the Duodenum', *Lancet*, 11 (1828), p. 320; J. Abercrombie, 'Case of Disease of the Duodenum', *Edinburgh Medical and Surgical Journal*, 44 (1835), pp. 277–8; J. Abercrombie, 'Perforating Ulcer of the Duodenum', *Edinburgh Medical and Surgical Journal*, 44 (1835), pp. 278–9; 'Case of Rupture of the Duodenum with Collapse and Peritonitis', *Lancet*, 43 (1844), pp. 23–4; 'The History of a Case of Chronic Ulceration of the Duodenum and Perforation', *Lancet*, 45 (1845), pp. 6–7; and G. Stilwell, 'Instance of Perforation of the Duodenum Occasioning Death after Twenty Hours' Illness', *Lancet*, 48 (1846), p. 67.
24. T. B. Curling, 'Acute Ulceration of the Duodenum in Cases of Burns', *Medico-Chirurgical Transactions*, 25 (1842), pp. 260–81, on p. 260. Further reports followed, including S. Gibbon, 'Ulcerated Duodenum Taken from a Man who Died in Consequence of an Extensive Scald', *Philosophical Transactions of the Royal Society of London*, 145 (1855), pp. 189–91.
25. Sonnenberg, 'Causes Underlying the Birth Cohort Phenomenon of Peptic Ulcer', p. 1091.
26. 'Hunger Pain and Duodenal Ulcer', *British Medical Journal*, 1 (1909), pp. 1036–7, on p. 1036.
27. B. Dawson, 'Discussion on the Pathogenesis, Diagnosis and Medical Treatment of Gastric Ulcer', *British Medical Journal*, 2 (1912), pp. 936–51, on p. 938.
28. 'Perforated Duodenal Ulcer: A Collective Report on a Series of 200 Cases of Perforated Duodenal Ulcer Treated in Edinburgh between 1896 and 1912', *Edinburgh Medical Journal*, 31 (1913), pp. 405–6.
29. J. Parr, *How I Cured My Duodenal Ulcer* (London: Michael Joseph, 1951).
30. Warren, 'Unidentified Curved Bacilli'.
31. G. D. Smith, 'The Biopsychosocial Approach: A Note of Caution', in P. White (ed.), *Biopsychosocial Medicine: An Integrated Approach to Understanding Illness* (Oxford: Oxford University Press, 2005), pp. 77–103.
32. M. D. Gershon, *The Second Brain: A Groundbreaking New Understanding of Nervous Disorders of the Stomach and Intestine* (New York and London: Harper Collins, 1998).
33. S. Levenstein, '"The Very Model of a Modern Etiology": A Biopsychosocial View of Peptic Ulcer', *Psychosomatic Medicine*, 62:3 (2000), pp. 176–85.
34. H. Spiro, 'Peptic Ulcer is not a Disease, Only a Sign! Stress is a Factor in More than a Few Dyspeptics', *Psychosomatic Medicine*, 62:2 (2000), pp. 186–7.

35. D. C. Delaney, M. Qume et al., 'Helicobacter Pylori Test and Treat versus Proton Pump Inhibitor in Initial Management of Dyspepsia in Primary Care: Multicentre Randomised Controlled Trial', *British Medical Journal*, 336 (2008), pp. 651–4.
36. Levenstein, 'Stress and Peptic Ulcer'.

WORKS CITED

Primary Sources

Abercrombie, J., *Pathological and Practical Researches on Diseases of the Stomach*, 2nd edn (Edinburgh: Waugh & Innes, 1830).

—, 'Case of Disease of the Duodenum', *Edinburgh Medical and Surgical Journal*, 44 (1835), pp. 277–8.

—, 'Perforating Ulcer of the Duodenum', *Edinburgh Medical and Surgical Journal*, 44 (1835), pp. 278–9.

Abernethy, J., *Surgical Observations on the Constitutional Origin and Treatment of Local Diseases* (London: Longman, Rees, Orme, Brown, Green & Longman, 1811).

—, *An Enquiry into the Probability and Rationality of Mr Hunter's Theory of Life* (London: Longman, 1814).

—, *The Abernethian Code of Health and Longevity* (London: J. Williams, 1829).

—, *Lectures on the Theory and Practice of Surgery* (London: Longman, Rees, Orme, Brown, & Green, 1830).

—, 'My Book by John Abernethy', *British Journal of Surgery*, 17:67 (1930), pp. 369–72.

Alexander, F., 'The Influence of Psychologic Factors upon Gastrointestinal Disturbances', *Psychoanalytic Quarterly*, 3 (1934), pp. 501–88.

—, *Psychosomatic Medicine* (New York and London: Norton, 1950).

Alexander, W. C., 'A Case Illustrating the Difficulty of Diagnosis in Gastric Ulcer', *British Medical Journal*, 1 (1897), p. 1345.

Alison, W. P., 'Observations on the Physiological Principle of Sympathy, Chiefly in Reference to the Doctrines of Mr Charles Bell', *Transactions of the Medico-Chirurgical Society of Edinburgh*, 2 (1826), pp. 165–226.

Allbutt, T. C., *On Visceral Neuroses* (London: J. Churchill, 1884).

Alvarez, W. C., *Nervous Indigestion* (London: William Heinemann, 1930).

Anon., 'The Alimentary Function and its Disorders', *Edinburgh Medical and Surgical Journal*, 17 (1821), pp. 574–608.

—, 'Dr Abercrombie on the Pathology of the Stomach, the Pancreas, and the Spleen', *Edinburgh Medical and Surgical Journal*, 21 (1824), pp. 1–15.

—, 'Dr Baillie's Posthumous Writings', *Medico-Chirurgical Review*, 4:8 (1826), pp. 364–79.

—, 'Mr Howship on Indigestion', *Medico-Chirurgical Review*, 4:7 (1826), pp. 34–43.

—, 'Diet in Britain and France', *Manchester Guardian and British Volunteer*, 14 July 1827, p. 4.

—, 'Uwins on Indigestion', *The Times*, 27 December 1827, p. 4.

—, 'Gastrology', *Medico-Chirurgical Review*, 9 (1828), pp. 217–18.

—, 'Ulceration of the Duodenum', *Lancet*, 11 (1828), p. 320.

—, 'Diseases of the Stomach', *Medico-Chirurgical Review*, 10 (1829), pp. 328–54.

—, 'Dr Abercrombie on the Diseases of the Stomach', *Edinburgh Medical and Surgical Journal*, 31 (1829), pp. 135–60.

—, 'Caution: Tight Stay Lacing', *Hull Packet and Humber Mercury*, 3 May 1831, p. 4.

—, 'A Word to Machine Markers', *Manchester Times and Gazette*, 14 January 1832, p. 2.

—, 'Treatment of Dyspepsia and Gastrodynia', *Dublin Journal of Medical and Chemical Science*, 2 (1832), pp. 143–4.

—, 'Dr Stokes on Chronic Gastritis and Dyspepsia', *Medico-Chirurgical Review*, 21 (1834), pp. 459–61.

—, 'Diet', *Derby Mercury*, 30 December 1835, p. 4.

—, 'Perforation of the Stomach Probably by a Worm', *Lancet*, 27 (1836), pp. 368–70.

—, 'Parker on the Stomach in its Morbid States', *Dublin Journal of Medical Science*, 13 (1838), pp. 334–49.

—, 'Death of Sir Francis Chantrey', *Manchester Guardian*, 1 December 1841, p. 1.

—, 'Post-Mortem Examination: A Drunkard's Body', *Temperance Lancet and Journal of Useful Intelligence*, 5:1 (1841), pp. 35–6.

—, 'A Short Sermon', *Manchester Times and Gazette*, 10 September 1842, p. 1.

—, 'Father Mathew at Kennington and Fulham', *Observer*, 13 August 1843, p. 4.

—, 'Fatal Effects of Tight Lacing', *Freeman's Journal and Daily Commercial Advertiser*, 15 April 1844, p. 4.

—, 'A Second Chapter on Taxation', *Manchester Guardian*, 24 April 1844, p. 6.

—, 'Case of Rupture of the Duodenum with Collapse and Peritonitis', *Lancet*, 43 (1844), pp. 23–4.

—, 'The History of a Case of Chronic Ulceration of the Duodenum and Perforation', *Lancet*, 45 (1845), pp. 6–7.

—, 'The Condition of the People of Ireland', *Freeman's Journal and Daily Commercial Advertiser*, 8 January 1846, p. 4.

—, 'Indigestion', *Lloyd's Weekly London Newspaper*, 27 September 1846, p. 5.

—, 'Present Condition of the Irish People', *Manchester Guardian*, 19 December 1846, p. 5.

—, 'Suicide of Professor M'Cullagh', *Freeman's Journal and Daily Commercial Advertiser*, 26 October 1847, p. 3.

—, 'Condiments and Sauces', *Truth Tester, Temperance Advocate and Healthian Journal*, 1 (1847), pp. 106–7.

—, 'Address to the Trades of England', *Northern Star and National Trades' Journal*, 3 June 1848, p. 1.

—, 'The Death-Bed of the Inebriate', *People's Abstinence Standard and True Social Reformer*, 1 (1849), p. 228.

—, 'Excessive Brandy Drinking: Perforation of the Stomach after Death', *Medical Times and Gazette*, 19 (1849), p. 242.

—, 'Chronic Ulcer of the Stomach', *American Journal of Medical Science*, 20 (1850), pp. 35–6.

—, 'Chronic Ulcer of the Stomach', *Philosophical Transactions of the Royal Society of London*, 140 (1850), pp. 211–12.

—, 'Temperance Festival in Blackburn', *Manchester Temperance Reporter and Journal of Progress*, 18:1 (1850), pp. 142–3.

—, 'Dyspepsia', *North Wales Chronicle*, 15 March 1851, p. 5.

—, 'Manchester, by a Manchester Man', *Preston Guardian*, 18 June 1853, p. 3.

—, 'Memoirs of John Abernethy, M. R. C. S.', *Observer*, 10 October 1853, p. 7.

—, 'A Case of Consumption', *Trewman's Exeter Flying Post*, 23 March 1854, p. 8.

—, 'A Cure of Consumption Induced by Indigestion, Exeter, after the Best of Routine Practice had Failed', *Trewman's Exeter Flying Post*, 20 April 1854, p. 4.

—, 'Struggles of the Spitalfields Poor', *Era*, 13 July 1856, p. 12.

—, 'The State of the Nation: A Word of Warning to the Working Classes', *Reynold's Newspaper*, 8 August 1858, p. 1.

—, 'Bread and Cheese v. Reform', *Lloyd's Weekly London Newspaper*, 12 December 1858, p. 1.

—, 'Meditations on Dyspepsia', *Blackwood's Edinburgh Review*, 90 (1861) p. 302.

—, 'Bayard on Diseases of the Stomach', *Dublin Quarterly Journal of Medical Science*, 34 (1862), pp. 342–9.

—, 'Drink and Insanity', *Glasgow Herald*, 18 August 1866, p. 2.

—, 'The Boy with Two Stomachs', *British Medical Journal*, 1 (1877), p. 669.

—, 'Reminiscences of an Old Bohemian', *The Times*, 25 October 1882, p. 4.

—, 'The Scientist at the Bedside', *Zoophilist*, 12 (1882), pp. 1–4.

—, 'The Experiences of M. De Lesseps', *Manchester Guardian*, 11 August 1883, p. 9.

—, 'Contribution to the Determination of the Percussion Limits of the Stomach', *Dublin Journal of Medical Science*, 3 (1887), pp. 21–40.

—, 'The Acids of the Stomach', *British Medical Journal*, 2 (1889), pp. 774–5.

—, 'The Effects of Extirpation of the Pancreas', *Lancet*, 136 (1890), p. 1407.

—, 'Perforation of Gastric Ulcer and its Treatment by Abdominal Section and Suture', *British Medical Journal*, 1 (1892), pp. 63–4.

—, 'Perigastric Abscess Perforating the Diaphragm and Causing Empyema; Operation; Death; Necropsy', *Lancet*, 142 (1893), pp. 89–91.

—, 'Operation in Perforation of Gastric Ulcer', *Practitioner*, 52 (1894), pp. 452–3.

—, 'The Scientific Use of Hospitals', *Nineteenth Century*, 49 (1901), pp. 57–63.

—, 'The Surgery of Chronic Dyspepsia', *Lancet*, 160 (1902), pp. 512–13.

—, 'A Discussion on the Surgical Treatment of the Non-Malignant Diseases of the Stomach', *British Medical Journal*, 2 (1905), pp. 767–810.

—, 'The Therapeutic Value of Hydrochloric Acid in Diseases of the Stomach', *British Medical Journal*, 2 (1905), pp. 891–7.

—, 'Recent Advances in the Surgery of the Stomach', *Lancet*, 167 (1906), pp. 609–10.

—, 'The Treatment of Gastric Ulcer', *British Medical Journal*, 1 (1908), pp. 387–91.

—, 'Forcible Feeding of Suffragists' *Manchester Guardian*, 29 September 1909, p. 4.

—, 'The Woman Suffragists', *The Times*, 29 September 1909, p. 10.

—, 'The Feeding of Suffragist Prisoners', *The Times*, 7 October 1909, p. 5.

—, 'Pried Open Teeth of Suffragettes' *New York Times*, 17 October 1909, p. C4.

—, 'A House Defaced and a Window Broken', *The Times*, 5 November 1909, p. 4.

—, 'Woman Suffrage', *The Times*, 12 November 1909, p. 12.

—, 'Woman Suffrage: The Fabian Society and Forcible Feeding', *The Times*, 19 November 1909, p. 10.

—, 'Fasting Prisoners and Compulsory Feeding', *British Medical Journal*, 2 (1909), pp. 997–8.

—, 'Forcible Feeding: Opinions of Medical Experts, Grave Danger to Life Involved', *Votes for Women*, 3:82 (1909), p. 3.

—, 'Hunger Pain and Duodenal Ulcer', *British Medical Journal*, 1 (1909), pp. 1036–7.

—, 'The Imprisonment of Lady Constance Lytton', *The Times*, 26 January 1910, p. 10.

—, 'Book Reviews', *Annals of Surgery*, 52:4 (1910), pp. 572–5.

—, 'Reviews and Bibliographical Notices', *Dublin Journal of Medical Science*, 131 (1910), pp. 132–3.

—, 'The Case of William Ball', *Manchester Guardian*, 10 May 1912, p. 4.

—, 'The Attempt at Suicide', *The Times*, 29 June 1912, p. 6.

—, 'Home Office Statements Refuted', *Votes for Women*, 5 (1912), p. 664.

—, 'A Physician on Forcible Feeding', *The Times*, 26 February 1913, p. 10.

—, 'Miss Lilian Lenton's Story of her Experience', *Manchester Guardian*, 22 October 1913, p. 10.

—, 'Against Forcible Feeding', *Suffragette*, 2 (1913), p. 194.

—, 'Perforated Duodenal Ulcer: A Collective Report on a Series of 200 Cases of Perforated Duodenal Ulcer Treated in Edinburgh between 1896 and 1912', *Edinburgh Medical Journal*, 31 (1913), pp. 405–6.

—, 'Suffragette who was Forcibly Fed 232 Times', *Manchester Guardian*, 17 April 1914, p. 9.

—, 'Notes and Notices', *Animals Defender and Zoophilist*, 35 (1915), p. 53.

—, 'The Law and the Women', *Common Cause*, 7 (1916), p. 570.

—, 'Psychiatric Casualties in London, September 1940', *British Medical Journal*, 2 (1940), pp. 553–5.

—, 'Total War and the Individual', *Lancet*, 237 (1941), pp. 791–2.

—, 'Recent Advances in Treatment', *Practitioner*, 151 (1943), pp. 258–62.

Ashe, T., *The Death of Thomas Ashe* (Dublin: J. M. Butler, 1917).

Bager, B., 'Beitral Zur Kenntnis über Vorkommen, Klinink und Behandlung von Perforierten Magenund Duodenalgeschuwüren Nebst Einer Untersuchung über die Spatresultate Nach Verschiedenen Operationsmethoden', *Acta Chirurgica Scandinavica*, 64 (1929), Supplement 11.

Baillie, M., *An Appendix to the First Edition of the Morbid Anatomy of Some of the Most Important Parts of the Human Body* (London: J. Johnson, 1798).

Ballatine, H., 'Remarks on Duodenal Ulcer with Notes of a Case', *Edinburgh Medical Journal*, 14 (1903), pp. 532–5.

Barclay, A. E., *The Stomach and Oesophagus: A Radiographic Study* (London: Sherratt & Hughes, 1913).

Barker, A. E., 'On Some Cases of Operation for Chronic Non-Malignant Disease of the Stomach', *Lancet*, 160 (1902), pp. 501–3.

Barling, G., 'The Treatment of Perforated Gastric Ulcer with Report of Successful Drainage in a Case', *British Medical Journal*, 1 (1893), pp. 1258–9.

—, 'An Address on the Diagnosis and Surgical Treatment of Certain Cases of Chronic Indigestion', *British Medical Journal*, 2 (1904), pp. 1623–7.

Barnes, D., 'How it Feels to be Forcibly Fed', *New York World Magazine*, 6 September 1914, pp. 5, 6, 17.

Bateman, D., *Berkeley Moynihan: Surgeon* (London: Macmillan & Co., 1940).

Beaumont, W., *Experiments and Observations on the Gatric Juice, and the Physiology of Digestion* (Plattsburgh, NY: F. P. Allen, 1833).

Benedetti, A., *Historia Corporis Humani* (1497).

Bennett, I., 'Peptic Ulcer in the Services', *British Medical Journal*, 1 (1940), p. 458.

Berdge, E., 'Torture of Animals in London Today', *Animals' Defender and Zoophilist*, 35 (1915), p. 15.

Bernard, C., *An Introduction to the Study of Experimental Medicine*, trans. H. C. Greene (New York: Macmillan, 1927).

Billroth, T., *Die Allgemeine Chirurgische Pathologie and Chirurgie* (Berlin: G. Reimer, 1863).

Bland-Sutton, J., *Essays on Hysterectomy* (London: Adlard, 1904).

Bolton, C., *Ulcer of the Stomach* (London: Edward Arnold, 1913).

—, 'Recent Observations on the Pathology of Ulcer of the Stomach with Indications for Treatment', *British Medical Journal*, 1 (1915), pp. 707–10.

Bolvin, M. A. V. G., and A. Dugès, *A Practical Treatise on the Diseases of the Uterus and its Appendages* (London: Sherwood, Gilbert & Piper, 1834).

Bosanquet, W. C., and H. S. Clogg, *The Stomach, Intestines and Pancreas* (London: John Bale, Sons & Danielsson Ltd, 1909).

Bourchier Nicholson, R. H., 'Case of Perforative Ulcer of Stomach Treated by Laparotomy, Stitching and Washing Out', *British Medical Journal*, 2 (1894), pp. 982–3.

Bradshaw, W., *The Anatomy of Dyspepsia* (London: George Philip & Sons, 1864).

—, *Brain and Stomach or Mind and Matter* (London: W. Philip & Son, 1867).

Bridger, A. E., *The Demon of Dyspepsia* (London: Swan Sonnenschein, Lowrey & Co., 1888).

Brinton, W., *On the Pathology, Symptoms and Treatment of Ulcer of the Stomach* (London: John Churchill, 1857).

Brockbank, W., 'The Dyspeptic Soldier: A Record of 931 Consecutive Cases', *Lancet*, 239 (1942), pp. 39–42.

Broussais, F. J. V., *A Treatise on Physiology Applied to Pathology*, trans. J. Bell and R. La Roche (Philadelphia, PA: H. C. Carey & I. Lea, 1826).

Brown, F. R., 'Duodenal Ulcer Battalions', *British Medical Journal*, 2 (1942), p. 530.

Brunton, T. L., *On Disorders of Digestion, their Consequences and Treatment* (London: Macmillan & Co., 1886).

Budd, G., *On the Organic Diseases and Functional Disorders of the Stomach* (London: J. Churchill, 1855).

Burne, P., *The Teetotaller's Companion* (London: Arthur Hall & Co., 1847).

Burns, A., 'Observations on the Digestion of the Stomach after Death', *Edinburgh Medical and Surgical Journal*, 6 (1810), pp. 129–38.

Cannon, W. B., 'The Influence of Emotional States on the Functions of the Alimentary Canal', *American Journal of Medical Science*, 137:4 (1909), pp. 480–6.

—, *The Mechanical Factors of Digestion* (London: Edward Arnold, 1911).

—, *Bodily Changes in Pain, Hunger, Fear and Rage: An Account of Recent Researches* (New York and London: D. Appleton & Co., 1915).

—, *Traumatic Shock* (New York: D. Appleton & Co., 1923).

—, *The Wisdom of the Body* (New York: Norton, 1932).

Carless, A., 'The Surgery of the Stomach', *Practitioner*, 12 (1900), pp. 491–501.

Carswell, R., 'An Inquiry on the Chemical Solution or Digestion of the Coats of the Stomach after Death', *Edinburgh Medical and Surgical Journal*, 34 (1830), pp. 282–311.

Chapman, J., *Functional Diseases of the Stomach Part One: Seasickness, its Nature and Treatment* (London: Trubner & Co., 1864).

Chevers, N., 'An Inquiry into Certain of the Causes of Death after Injuries and Surgical Operations in London Hospitals', *Guy's Hospital Reports*, 2 (1843), pp. 78–101.

Cheyne, G., *The English Malady* (London, 1733).

Clarke, J. E., *Indigestion: Its Causes and Cures* (London: James Epps & Co., 1888).

Conwell, W. E. E., *A Treatise on the Functional and Structural Changes of the Liver* (London: James Duncan, 1835).

Coope, R., 'Recent Trends in Gastro-Enterological Treatment', *Practitioner*, 149:5 (1942), pp. 277–83.

Cooper, A., and B. Travers, *Surgical Essays: Part One* (London: Cox, 1818).

Cooper, J. C., 'Description of the Operation of Gastrotomy', *Guy's Hospital Reports*, 4 (1858), pp. 13–18.

Cullen, W., *First Lines of the Practice of Physic* (Edinburgh: William Creech, 1777).

Culverwell, R. J., *Hints to the Nervous and Dyspeptic on the Causes and Cure of Nervousness, Indigestion, Haemorrhoids and Constipation* (London: for the author, 1837).

Curling, T. B., 'Acute Ulceration of the Duodenum in Cases of Burns', *Medico-Chirurgical Transactions*, 25 (1842), pp. 260–81.

Cushing, H., 'Peptic Ulcer and the Interbrain', *Surgery, Gynaecology and Obstetrics*, 55:1 (1932), pp. 1–34.

Dalziel, T. K., 'Thirty Cases of Gastro-Enterostomy for Non-Malignant Affections of the Stomach', *Lancet*, 160 (1902), pp. 503–4.

Davies, D. T., and A. T. Macbeth Wilson, 'Observations on the Life-History of Chronic Peptic Ulcer', *Lancet*, 230 (1937), pp. 1353–60.

Dawson, B., 'The Diagnosis of Gastric Ulcer', *British Medical Journal*, 2 (1905), pp. 1032–4.

—, 'Discussion on the Pathogenesis, Diagnosis and Medical Treatment of Gastric Ulcer', *British Medical Journal*, 2 (1912), pp. 936–51.

Dewar, J., *The Red Cross Series of Health Hand Books: Dyspepsia* (London: William Patterson & Co., 1891).

Dobson, N. C., 'Abdominal Section in Perforating Ulcer of the Stomach', *Bristol Medico-Chirurgical Review*, 7 (1883), p. 196.

Douthwaite, A. H., 'Gastro-Enterology', *Practitioner*, 147 (1941), pp. 622–9.

Drewry, G. O., *Common-Sense Management of the Stomach* (London: Henry S. King & Co., 1875).

Druitt, R., *The Principles and Practice of Modern Surgery*, 3rd edn (Philadelphia, PA: Blanchard & Lea, 1852).

Durey, M. J., 'Bodysnatchers and Benthamites: The Implications of the Dead Body Bill for the London Schools of Anatomy 1820–42', *London Journal*, 2 (1976), pp. 200–25.

Einhorn, M., *Diseases of the Stomach* (London: Baillière & Co., 1897).

Elder, G., *Three Cases of Nephrectomy with Remarks on the Operation* (London: John Bale & Sons, 1885).

Eno, J. C., *The Stomach and its Trials* (London: F. Newbury & Sons, 1871).

Erichsen, J. E., *The Science and Art of Surgery being a Treatise on Surgical Injuries, Diseases and Operations*, 5th edn (London: James Walton, 1869).

Ewald, C. A., *Diseases of the Stomach* (Edinburgh: Young J. Pentland, 1892).

Fenwick, S., *The Morbid States of the Stomach and Duodenum and their Relation to the Diseases of other Organs* (London: John Churchill & Sons, 1868).

Fenwick, W. S., *Dyspepsia: Its Varieties and Treatment* (Philadelphia, PA, and London: W. B. Saunders, 1910).

Fitz, R. H., 'Acute Pancreatitis', *Boston Medical and Surgical Journal*, 120:8 (1889), pp. 181–7.

Friedenwald, J., 'Note on the Discovery and Early Use of the Stomach Tube', *Medical Life*, 34:12 (1927), pp. 639–44.

—, 'The History of the Development of the Stomach Tube with Some Notes on the Duodenal Tube', *Bulletin of the History of Medicine*, 4:6 (1936), pp. 425–54.

Gibbon, S., 'Ulcerated Duodenum Taken from a Man who Died in Consequence of an Extensive Scald', *Philosophical Transactions of the Royal Society of London*, 145 (1855), pp. 189–91.

Gilford, H., 'A Case of Perforated Gastric Ulcer for which Gastrorrhaphy was Performed: Death on the Thirty-First Day', *British Medical Journal*, 1 (1893), pp. 944–6.

—, 'The Surgical Treatment of (Unperforated) Gastric Ulcer with an Account of Three Cases in which Operations were Performed', *Guy's Hospital Reports*, 53 (1898), pp. 103–25.

Gillespie, A. L., *A Manual of Modern Gastric Methods: Chemical, Physical and Therapeutical* (Edinburgh: Oliver & Boyd, 1899).

Gilsenan, B., 'Dyspepsia of Peptic Ulcer Type and its Relationship to Personality Type and Anxiety', *Practitioner*, 157:2 (1946), pp. 457–8.

Gmelin, L., and F. Tiedemann, *Die Verdauung nach Versuchen* (Heidelberg: Karl Gros, 1827).

Graham, J. G., and J. D. O. Kerr, 'Digestive Disorders in the Force', *British Medical Journal*, 1 (1941), pp. 473–6.

Graham, T. J., *A Treatise on Indigestion*, 3rd edn (London: Simpkin & Marshall, 1833).

Griffiths, P. R., 'Perforating Gastric Ulcer with Notes of Two Successful Cases after Operation', *British Medical Journal*, 1 (1900), pp. 572–4.

Hageby, L. L., and L. K. Schartau, *The Shambles of Science* (London: Ernest Bell, 1903).

Hall, C., *The Effects of Civilisation on European States* (1805; Cambridge: Chadwyck-Healey Ltd, 1995).

Hare, H. A., and E. Q. Thornton, 'A Study of the Influence of Chloroform upon the Respiration and Circulation', *Lancet*, 142 (1893), pp. 996–9.

Harley, V., 'The Normal Absorption of Fat and the Effect of Extirpation of the Pancreas on It', *Journal of Physiology*, 18:1 (1895), pp. 1–14.

—, *Pancreatic Diabetes in Animals and Man* (Manchester, 1895).

Harvey, W., *Exercitatio Anatomica de Motu Cordis et Sanguinis in Animalibus* (Francofurti: Sumptibus G. Fitzeri, 1628).

Herschell, G., *Manual of Intragastric Technique* (London: H. J. Glaisher, 1903).

—, 'A New Gastro-Diaphane', *Lancet*, 163 (1904), p. 1361.

—, 'An Improved Apparatus for Auto-Lavage of the Stomach', *Lancet*, 164 (1904), p. 532.

—, *The Surgical Treatment of Duodenal Ulcer* (London: Henry J. Glashier, 1910).

Hill, A. B., 'An Investigation into the Sickness Experience of London Transport Workers, with Special Reference to Digestive Disturbances', *Medical Research Council International Health Research Board Report* (London: His Majesty's Stationary Office, 1937).

Hill, W., *On Gastroscopy and Oesophago-Gastroscopy* (London: John Bale, 1912).

Hinds-Howell, C. A., 'A Review of Dyspepsia in the Army', *British Medical Journal*, 2 (1941), pp. 473–4.

Hodson, V. S., 'Duodenal Ulcers and Cigarette Smoking', *Lancet*, 228 (1936), pp. 1235–6.

Hope, J., *A Treatise on the Diseases of the Heart and Great Vessels* (London: John Churchill, 1832).

Horsley, V., 'Forcible Feeding', *The Times*, 21 December 1909, p. 10.

Horwich, J. J., 'Incidence of Peptic Ulcer', *British Medical Journal*, 2 (1944), p. 866.

Howship, J., *Practical Remarks upon Indigestion* (London: Longman, Hurst, Rees, Orme, Brown, & Green, 1825).

Hughes, H. M., 'Case of Supposed Spontaneous Perforation of the Stomach', *Guy's Hospital Reports*, 4 (1846), pp. 332–42.

Hunt, T. C., *The Common Causes of Chronic Indigestion: Differential Diagnosis and Treatment* (London: Balliere, Tindall & Cox, 1933).

Hunter, J., 'On the Digestion of the Stomach after Death', *Philosophical Transactions of the Royal Society of London*, 62 (1772), pp. 447–54.

—, *A Treatise on the Blood, Inflammation and Gun-Shot Wounds* (London: John Richardson, 1794).

—, *The Works of John Hunter F.R.S.*, ed. J. F. Palmer, 4 vols (London: Longman, Rees, Orme, Brown, Green & Longman, 1837).

Hurst, A., 'New Views on the Pathology, Diagnosis and Treatment of Gastric and Duodenal Ulcer', *British Medical Journal*, 1 (1920), pp. 559–63.

—, 'Duodenal Ulcer and the Hypertonic Stomach', *British Medical Journal*, 1 (1920), p. 688.

—, 'Wartime Diet for Peptic Ulcer Patients', *British Medical Journal*, 2 (1943), p. 523.

—, *Medical Diseases of War*, 4th edn (London: Edward Arnold & Co., 1944).

Hurst, A. F., and M. J. Stewart, *Gastric and Duodenal Ulcer* (London and New York: H. Milford, Oxford University Press, 1929).

Hyams, K. C., F. S. Wignall and R. Roswell, 'War Syndromes and their Evaluation: From the US Civil War to the Persian Gulf War', *Annals of Internal Medicine*, 125:5 (1996), pp. 398–405.

Illingworth, C. F. W., and L. D. W. Scott, 'Acute Perforated Peptic Ulcer: Frequency and Incidence in the West of Scotland', *British Medical Journal*, 2 (1944), pp. 617–18.

Jackson, J. C., 'Tired Stomachs', *Dietetic Reformer and Vegetarian Messenger*, 79 (1881), p. 152.

James, G. W. G., 'Anxiety Neurosis', *Lancet*, 236 (1940), pp. 561–4.

Jennings, F. D., 'Perforated Peptic Ulcer: Changes in Age-Incidence and Sex-Distribution in the Last 150 Years', *Lancet*, 235 (1940), pp. 395–8, 444–7.

Jessett, F. B., *Surgical Diseases and Injuries of the Stomach and Intestines* (London: Baillière & Co., 1892).

Johnson, J., *The Influence of Tropical Climates* (London, 1813).

—, *An Essay on Morbid Sensibility of the Stomach and Bowels, as the Proximate Cause of Indigestion* (London, 1827).

Jordan, A. C., 'Duodenal Obstruction as Shown by Radiography', *British Medical Journal*, 1 (1911), pp. 1172–4.

Leared, A., *The Causes and Treatment of Imperfect Digestion* (London: John Churchill, 1860).

Lee, R., 'Analysis and Summary of One Hundred and Sixty Two Cases of Ovariotomy or Excision of the Ovarium which have Taken Place in Great Britain', *Edinburgh Medical and Surgical Journal*, 78 (1852), pp. 70–95.

Lewis, A., 'The Psychological Aspects of Indigestion', *Practitioner*, 153:4 (1944), pp. 257–60.

Lovell, R., *A Compleat History of Animals and Minerals* (Oxford: Goodwin, 1661).

Lytton, C., *Prisons and Prisoners: Some Personal Experiences* (London: Heinemann, 1914).

Maclean, H., *Modern Views on Digestion and Gastric Disease* (London: Constable & Co. Ltd, 1925).

—, 'A Lecture on the Intensive Alkaline Treatment of Gastric and Duodenal Ulcer', *British Medical Journal*, 1 (1928), pp. 619–23.

Mailer, R., 'The Later Results of Operations on the Stomach and Duodenum', *Glasgow Medical Journal*, 108 (1936), p. 110.

Major, R. H., 'History of the Stomach Tube', *Annals of Medical History*, 6 (1934), pp. 500–9.

Malcolm, J. D., *The Physiology of Death from Traumatic Fever: A Study in Abdominal Surgery* (London: J. & A. Churchill, 1893).

Mansell-Moullin, C., 'To the Editor of the Times', *The Times*, 29 September 1909, p. 10.

Marmaduke Shield, A., 'Two Cases of Ulcer of the Duodenum in which Laparotomy was Performed with Remarks on Ulcers of the Duodenum', *Lancet*, 145 (1895), pp. 1169–74.

Maylard, A. E., *A Treatise on the Surgery of the Alimentary Canal: Comprising the Oesophagus, the Stomach, the Small and Large Intestines, and the Rectum* (London: J. & A. Churchill, 1896).

—, 'An Address on the Surgery of the Stomach with Illustrative Cases', *Lancet*, 156 (1900), pp. 1253–8.

Mayo, T., *An Essay on the Influence of Temperament in Modifying Dyspepsia by Indigestion* (London, 1831).

Mayo-Robson, A. W., 'Abstracts of the Hunterian Lectures on the Surgery of the Stomach', *British Medical Journal*, 1 (1900), pp. 562–6.

—, 'Abstracts of the Hunterian Lectures on the Surgery of the Stomach II', *British Medical Journal*, 1 (1900), pp. 624–9.

—, 'The Hunterian Lecture on Duodenal Ulcer and its Treatment', *British Medical Journal*, 1 (1907), pp. 248–54.

Mayo-Robson, A. W., and B. Moynihan, *Diseases of the Pancreas and their Surgical Treatment* (Philadelphia, PA, and London: W. B. Saunders & Co., 1902).

—, *Diseases of the Stomach and their Surgical Treatment*, 2nd edn (London: Ballière, Tindall & Cox, 1904).

Middleton, G. S., 'Some Observations on the Difficulties of Diagnosis in Abdominal Disease, with Brief Accounts of Illustrative Cases', *Glasgow Medical Journal*, 5 (1897), pp. 322–39.

Mitchell, A. B., 'A Contribution to the Surgery of Perforated Gastric Ulcer', *British Medical Journal*, 1 (1900), pp. 567–75.

—, 'Contribution to a Discussion on the Treatment of Gastric Ulcer from a Surgical Standpoint', *Dublin Journal of Medical Science*, 123 (1907), pp. 179–80.

Morgan, W. G., 'Some Experiences with the Einhorn Duodenal Bucket and a Modified Thread Test', *American Journal of Medical Science*, 141:5 (1911), pp. 649–58.

Morris, J. N., and R. M. Titmuss, 'Epidemiology of Peptic Ulcer Vital Statistics', *Lancet*, 244 (1944), pp. 841–5.

Morton, W. T. G., *Remarks on the Proper Mode of Administering Sulphuric Ether by Inhalation* (Boston, MA, 1847).

Moxey, D. A., 'Feeding by the Nose in Attempted Suicide by Starvation', *Lancet*, 100 (1872), pp. 444–6.

—, 'Feeding by the Nose in Attempted Suicide by Starvation cont...', *Lancet*, 100 (1872), pp. 489–90.

Moynihan, B., 'On Total Extirpation of the Stomach with a Record of an Unsuccessful Case', *British Medical Journal*, 2 (1903), pp. 1458–9.

—, 'Duodenal Ulcer', *Practitioner*, 78 (1907), pp. 747–61.

—, *Duodenal Ulcer* (Philadelphia, PA, and London: W. B. Saunders Co., 1910).

—, *The Pathology of the Living and Other Essays* (Philadelphia, PA, and London: W. B. Saunders Co., 1910).

—, 'A Hunterian Lecture on Some Problems of Gastric and Duodenal Ulcer', *British Medical Journal*, 1 (1923), pp. 221–6.

—, *Two Lectures on Gastric and Duodenal Ulcer* (Bristol: John Wright & Sons, 1923).

—, 'On the Recognition of some Acute Abdominal Diseases', *Practitioner*, 63 (1931), p. 7.

—, *The Advance of Medicine* (Oxford: Clarendon Press, 1932).

Munro, A., *The Morbid Anatomy of the Human Gullet, Stomach and Intestines* (Edinburgh: A. Constable & Co., 1811).

Newman, D., *Lectures to Practitioners on the Diseases of the Kidney Amenable to Surgical Treatment* (London and New York: Longmans, Green & Co., 1888).

Niles, G. M., *The Diagnosis and Treatment of Digestive Diseases* (London: Henry Kimpton, 1914).

Ord, W. M., 'On the Diagnosis and Treatment of Gastric Ulcer', *Dublin Journal of Medical Science*, 88 (1889), pp. 545–62.

Parker, L., *The Stomach in its Morbid States* (London: Longman, Orme, Brown, Green, 1838).

Parr, J., *How I Cured My Duodenal Ulcer* (London: Michael Joseph, 1951).

Parsons, A. R., 'The Desirability of Operative Interference in Suspected Perforation of Chronic Ulcer of the Stomach', *Dublin Journal of Medical Science*, 94 (1892), pp. 26–35.

—, 'The Diagnosis and Treatment of a Perforated Gastric Ulcer, with Notes on a Successful Case', *Dublin Journal of Medical Science*, 121 (1906), pp. 81–92.

Parsons, B., and J. Marsh, *Anti-Bacchus: An Essay on the Crime, Diseases and other Evils Connected with the Use of Intoxicating Drinks* (London: John Snow, 1840).

Payne, R. T., and C. Newman, 'Interim Report on Dyspepsia in the Army', *British Medical Journal*, 2 (1940), pp. 619–21.

Pegge, G., 'Psychiatric Casualties in First Days of War', *British Medical Journal*, 2 (1939), pp. 764–5.

Perrengton, S., *A Popular Treatise on the Stomach with Certain New and Important Principles in the Treatment of Indigestion* (London: Charles & Edwin Layton, 1839).

Perry, C., and L. E. Shaw, 'Malignant Disease of the Stomach', *Guy's Hospital Reports*, 58 (1904), pp. 121–52.

Pethick Lawrence, F. W., 'Newcastle Prisoners Released: The Story of their Ordeal in Prison', *Votes for Women*, 3 (1909), p. 67.

Philip, A. P. W., *A Treatise on Indigestion and its Consequences*, 6th edn (London: Thomas & George Underwood, 1828).

Pirie, G. A., 'The Diagnosis of Disease of the Stomach by the X-Rays', *Edinburgh Medical Journal*, 9 (1912), pp. 137–42.

Prout, W., *Chemistry, Meteorology and the Function of Digestion Considered with Reference to Natural Theology* (London: W. Pickering, 1834).

Purser, J. M., 'On the Modern Diagnosis of Diseases of the Stomach', *Dublin Journal of Medical Science*, 90:6 (1890), pp. 449–87.

Ray, E., 'Case of Ulcer in the Stomach Leading to Perforation and Death in Nineteen Hours', *Guy's Hospital Reports*, 4 (1846), pp. 343–50.

Réaumur, R., 'Sur la Digestion des Oiseaux', *Memoires de l'Academie Royale des Sciences*, 1756 (1752), pp. 266–307.

Rees, G., *Practical Observations on Disorders of the Stomach*, 2nd edn (London: M Allen, 1811).

Reeves, C. E., *Diseases of the Stomach and Duodenum* (London: Simpkin, Marshall, 1856).

Rehfuss, M. E., *Indigestion: Its Diagnosis and Management* (Philadelphia, PA, and London: W. B. Saunders Co., 1943).

Reid, J., *Essays on Insanity, Hypochondriasis and other Nervous Affections* (London: Longman, Hurst, Rees, Orme, & Brown, 1816).

Riegel, F., 'Diagnosis of Gastric Disease', *Lancet*, 147 (1896), p. 568.

—, *Diseases of the Stomach* (Philadelphia, PA, and London: W. B. Saunders, 1903).

Roberts, W., *Lectures on Dietetics and Dyspepsia* (London: Smith, Elder, 1885).

Romanis, W. H. C., 'The Surgical Aspect of Gastric Ulcer', *Practitioner*, 55 (1923), p. 232.

Rowe, J. B. W., 'Wartime Diet for Peptic Ulcer Patients', *British Medical Journal*, 2 (1943), p. 464.

Sampson, O. E. I., 'Chronic Duodenal Ulcer and Perforation', *British Medical Journal*, 2 (1944), pp. 864–9.

Saundby, R., 'A Clinical Lecture on Ulcer of the Stomach', *British Medical Journal*, 1 (1900), pp. 121–2.

—, 'Remarks on the Indications for Operative Interference in Diseases of the Stomach', *British Medical Journal*, 2 (1904), pp. 1621–3.

—, *The Treatment of Diseases of the Digestive System* (London: Charles Griffin & Co., 1906).

Schlatter, C., 'Oesophagy-Enterostomy after Total Extirpation of the Stomach', *Lancet*, 151 (1898), pp. 141–6.

—, 'Further Observations on a Case of Total Extirpation of the Stomach in the Human Subject', *Lancet*, 152 (1898), pp. 1314–18.

Scott, R. A. M., 'Incidence of Peptic Ulcer', *British Medical Journal*, 1 (1945), p. 457.

Seyle, H., 'Perforated Peptic Ulcer during Air-Raid', *Lancet*, 241 (1943), p. 252.

Sippy, B. W., 'Gastric and Duodenal Ulcer: Medical Cure by an Efficient Removal of Gastric Juice Corrosion', *Journal of the American Medical Association*, 64 (1915), pp. 1625–30.

Simpson, J. Y., *Ovariotomy: Is it or is it not an Operation Justifiable upon the Common Principles of Surgery?* (Edinburgh: Sutherland & Knox, 1846).

Sinclair, W. J., 'Vaginal Hysterectomy for Cancer', *Practitioner*, 43:6 (1889), pp. 413–34.

Smith, J. G., *Abdominal Surgery* (London: John Churchill, 1891).

Smyth, M. J., 'Gastro-Jejunal Ulcer', *Irish Journal of Medical Science*, 9 (1931), p. 552.

Society for Protection of Animals from Vivisection, 'Scientific Medicine', *Society for Protection of Animals from Vivisection Pamphlets 2* (London: Office of the Society for Protection of Animals from Vivisection, 1881–2), pp. 1–2.

—, 'Samples of Vivisection', *Society for Protection of Animals from Vivisection Pamphlets* (London: Office of the Society for Protection of Animals from Vivisection, 1895), p. 1.

—, 'Samples of Vivisection', *Society for Protection of Animals from Vivisection Pamphlets* (London: Office of the Society for Protection of Animals from Vivisection, 1896), p. 2.

—, 'Some 1896 Vivisections', *Society for Protection of Animals from Vivisection Pamphlets* (London: Office of the Society for Protection of Animals from Vivisection, 1897), pp. 1–2.

Southam, G., *Ovariotomy: Removal of an Encysted Tumour of the Left Uterine Appendages* (Salford: W. F. Jackson, 1845).

Southwell, T., *Medical Essays and Observations* (London: J. Knox, 1764).

Souttar, H. S., and T. Thompson, 'The Direct Inspection of the Gastric Mucous Membrane', *Quarterly Journal of Medicine*, 1:4 (1908), pp. 376–9.

Spallanzani, L., *Dissertations Relative to the Natural History of Animals and Vegetables* (London: John Murray, 1784).

Speer, T. C., *General Views Relating to the Stomach, its Fabric and Functions* (London: Longman, Hurst & Rees, 1818).

Spencer Wells, T., *Modern Abdominal Surgery* (London: J. & A. Churchill, 1891).

Spicer, C. C., D. N. Stewart and D. M. de R. Winser, 'Incidence of Perforated Peptic Ulcer: Effect of Heavy Air Raids', *Lancet*, 239 (1942), pp. 259–61.

—, 'Perforated Peptic Ulcer during the Period of Heavy Air-Raids', *Lancet*, 243 (1944), p. 14.

Starling, E. H., *Recent Advances in the Physiology of Digestion* (London: Archibald Constable & Co., 1906).

Steven, J. L., 'On the Surgical Treatment of Diseases of the Stomach from a Physician's Point of View', *Lancet*, 163 (1904), pp. 1487–93.

Stevens, E., *Dissertatio Physiologica Inauguralis de Alimentarium Concoctione* (Edinburgh: Balfour & Smellie, 1777).

Stilwell, G., 'Instance of Perforation of the Duodenum Occasioning Death after Twenty Hours' Illness', *Lancet*, 48 (1846), p. 67.

Stone, A. D., *A Practical Treatise on the Disease of the Stomach* (London: Cadell & Davies, 1806).

Tait, L., *Diseases of Women and Abdominal Surgery* (Leicester: Richardson & Co., 1889).

Talamon, C. H., *Appendicitis and Perityphlitis* (Edinburgh and London: Young J. Pentland, 1893).

Taylor, A. S., 'On Perforations of the Stomach from Poisoning and Disease', *Guy's Hospital Reports*, 5 (1839), pp. 8–62.

—, *A Manual of Medical Jurisprudence* (London: John Churchill, 1844).

Taylor, F., 'Examination of Stomach by X-Rays and Tube Filled with Subnitrate of Bismuth', *British Medical Journal*, 1 (1905), p. 720.

Taylor, G. G., 'The Abdominal Surgery of Total War', *Glasgow Medical Journal*, 19:6 (1942), pp. 123–42.

Thomas, K., *Contributions to the Surgical Treatment of Tumours of the Abdomen* (Edinburgh: Oliver & Boyd, 1885).

Thornton, J. K., *Cases of Hysterectomy: With Remarks on the Value of the Carbolic Acid Spray in this Operation* (London: John Bale & Sons, 1885).

Tidy, H., 'Peptic Ulcer and Dyspepsia in the Army', *British Medical Journal*, 2 (1943), pp. 473–7.

—, 'Peptic Ulcer', *Practitioner*, 153:4 (1944), pp. 197–203.

Travers, B., 'Additional Observations on a Report of Rupture of the Stomach and Escape into the Cavity of the Abdomen by Crampton', *Medico-Chirurgical Transactions*, 8:1 (1817), pp. 228–45.

Titmuss, R. M., *Problems of Social Policy* (London: His Majesty's Stationary Office; Longmans, Green & Co., 1950).

Trotter, T., *A View of the Nervous Temperament* (London: Longman, Hurst, Rees & Orme, 1807).

Trowell, O. A., 'The Relation of Tobacco Smoking to the Incidence of Chronic Duodenal Ulcer', *Lancet*, 223 (1934), pp. 808–9.

Twigg, F., 'Perforation of Duodenal Ulcer While Flying', *British Medical Journal*, 2 (1930), p. 687.

Valzah, W. V., and J. D. Nisbet, *The Diseases of the Stomach* (London: Rebman Publishing Co., 1899).

Verity, R., *Changes Produced in the Nervous System by Civilisation* (London: S. Highley, 1837).

Warren, G., *A Discourse upon National Dietetics, as Connected with Dyspepsia, Hypochondriasis, Gout, Erysipelas, and many Diseases of this and other Christian Countries* (London, 1830).

Wheeler, I. W., 'The Diagnosis and Treatment of Gastric and Duodenal Ulcers', *Irish Journal of Medical Science*, 1 (1923), pp. 1–12.

Whiting, S., *Memoirs of a Stomach* (London: W. E. Painter, 1853).

—, *Mémoires d'un estomac*, trans. C. H. Gros (Paris: J. B. Baillière, 1888).

Whytt, R., *Observations on the Nature, Cause and Cure of those Disorders which have been commonly called Nervous Hypochondriac or Hysteric* (Edinburgh: J. Balfour, 1765).

Wightman, C., *A Treatise on the Sympathetic Relationship between the Stomach and the Brain* (London: Simpkin, Marshall & Co., 1840).

Wilcox, P. H., 'Gastric Disorders in the Services', *British Medical Journal*, 1 (1940), pp. 1008–12.

Williams, C., *Carcinoma of the Kidney, Nephrectomy: A Case of Gastrostomy for Malignant Disease of the Oesophagus* (London, 1892).

Williamson, T., 'An Attempt to Estimate some of the Characteristic Marks by which to Judge of the Cause of Perforations of the Stomach II', *Dublin Journal of Medical Science*, 19 (1841), pp. 191–219.

Wollaston, W. H., 'The Croonian Lecture', *Philosophical Transactions of the Royal Society of London*, 100:1 (1810), pp. 1–15.

Woodforde, J., *A Treatise on Dyspepsia or Indigestion*, 2nd edn (London: Longman, Hurst, Rees, Orme & Brown, 1821).

Wright, A., *Digestion: The Passport to Health* (London: William Rider & Son, 1883).

Yeats, G. D., *Some Observations on the Duodenum or Second Stomach* (London: G. Woodfall, 1820).

Young, E., *On Perforating Ulcer of the Stomach* (London: Simpkin, Marshall, & Co., 1849).

Secondary Sources

Ackerknecht, E., 'Broussais or a Forgotten Medical Revolution', *Bulletin of the History of Medicine*, 27 (1953), pp. 320–43.

—, *Medicine at the Paris Hospital 1794–1848* (Baltimore, MD: Johns Hopkins University Press, 1967).

—, *Therapeutics: From the Primitives to the Twentieth Century* (London: Macmillan, 1973).

Akiyama, Y., *Feeding the Nation: Nutrition and Health in Britain before World War One* (London: Tauris Academic Studies, 2008).

Albaba, K., *Eating Right in the Renaissance* (Berkeley, CA: University of California Press, 2002).

Alberti, F. B, 'Angina Pectoris and the Arnolds: Emotions and Heart Disease in the Nineteenth Century', *Medical History*, 52:2 (2008), pp. 221–36.

—, *Matters of the Heart: History, Medicine and Emotion* (Oxford: Oxford University Press, 2010).

Anning, S. T., 'The Practice of Surgery in Leeds 1823–1824', *Medical History*, 23:1 (1979), pp. 59–95.

Arrizabalaga, J., 'Medical Causes of Death in Preindustrial Europe: Some Historiographical Considerations', *Journal of the History of Medicine and Allied Sciences*, 54:2 (1999), pp. 241–60.

—, 'Problematising Retrospective Diagnosis in the History of Disease', *Asclepio*, 54:1 (2002), fasc. 1, pp. 51–70.

Astrup, P., *Salt and Water in Culture and Medicine* (Copenhagen: Munksgaard, 1993).

Babington, A., *Shell-Shock: A History of the Changing Attitudes to War Neurosis* (Barnsley: Cooper, 1997).

Baron, J. H., 'Peptic Ulcer', *Mount Sinai Journal of Medicine*, 67:1 (2000), pp. 58–62.

—, 'Alimentary Diseases in the Poor and Middle Class in London 1773–1815 and in New York Poor 1797–1818', *Alimentary Pharmacology and Therapeutics*, 16:10 (2002), pp. 1709–14.

—, 'Hospital Admissions for Peptic Ulcer and Indigestion in London and New York in the 19th and 20th Centuries', *Gut*, 50:4 (2002), pp. 568–70.

—, 'Publications on Peptic Ulcer in Britain, France, Germany and the US', *European Journal of Gastroenterology and Hepatology*, 214:7 (2002), pp. 711–15.

Baron, J. H., and A. Sonnenberg, 'Period- and Cohort-Age Contours of Death from Gastric and Duodenal Ulcer in New York 1804–1998', *American Journal of Gastroenterology*, 43:10 (2001), pp. 2887–91.

—, 'History of Dyspepsia in Scotland: Admissions to the Edinburgh Royal Infirmary 1729–1830', *Scottish Medical Journal*, 53:3 (2008), pp. 42–4.

Baron, J. H., F. Watson and A. Sonnenberg, 'Three Centuries of Stomach Symptoms in Scotland', *Alimentary Pharmacology and Therapeutics*, 24:5 (2006), pp. 821–9.

Bast, T. H., *The Life and Times of Adolf Kussmaul* (New York: P. B. Hoeber, 1926).

Bates, A. W., *Emblematic Monsters: Unnatural Conceptions and Deformed Births in Early Modern Europe* (Amsterdam and New York: Rodopi, 2005),

Bonner, T. N., *Becoming a Physician: Medical Education in Britain, France, Germany and the United States 1750–1945* (Oxford: Oxford University Press, 1995).

Brieger, G. H., 'Dyspepsia: The American Disease? Needs and Opportunities for Research', in C. E. Rosenberg (ed.), *Healing and History: Essays for George Rosen* (New York: Science History Publications, 1979), pp. 179–90.

—, 'From Conservative to Radical Surgery in Late Nineteenth-Century America', in Lawrence (ed.), *Medical Theory, Surgical Practice*, pp. 216–31.

Brock, W. H., *Justus von Liebig: The Chemical Gatekeeper* (Cambridge: Cambridge University Press, 1997).

Brown, M., 'Medicine, Reform and the "End" of Charity in Early Nineteenth-Century England', *English Historical Review*, 125:511 (2010), pp. 1353–88.

Brumberg, J. J., *Fasting Girls: The Emergence of Anorexia Nervosa as a Modern Disease* (Cambridge, MA: Harvard University Press, 1988).

Burney, I. A., *Bodies of Evidence: Medicine and the Politics of the English Inquest 1830–1926* (Baltimore, MD: Johns Hopkins University Press, 1999).

—, 'Medicine in the Age of Reform', in A. Burns and J. Innes (eds), *Rethinking the Age of Reform: Britain 1780–1850* (Cambridge: Cambridge University Press, 2003), pp. 163–81.

—, *Poison, Detection and the Victorian Imagination* (Manchester: Manchester University Press, 2006).

Busfield, J., *Managing Madness: Changing Ideas and Practices* (London: Ulman Hyman, 1986).

Bynum, W. F., *Science and the Practice of Medicine in the Nineteenth Century* (Cambridge: Cambridge University Press, 1994).

— (ed.), *Gastroenterology in Britain: Historical Essays* (London: Wellcome Institute for the History of Medicine, 1997).

Bynum, W. F., and J. C. Wilson, 'Periodical Knowledge: Medical Journals and their Editors in Nineteenth-Century Britain', in W. F. Bynum, S. Lock, and R. Porter (eds), *Medical Journals and Medical Knowledge: Historical Essays* (London: Routledge, 1992), pp. 9–48.

Canguilhem, G., *The Normal and the Pathological* (New York: Zone Books, 1989).

Cantor, D., 'The Diseased Body', in R. Cooter and J. V. Pickstone (eds), *Medicine in the Twentieth Century* (Amsterdam: Harwood Academic, 2000), pp. 347–64.

— (ed.), *Reinventing Hippocrates* (Aldershot: Ashgate, 2001).

Carden-Coyne, A., 'American Guts and Military Manhood', in Carden-Coyne and Forth (eds), *Cultures of the Abdomen*, pp. 71–86.

Carden-Coyne, A., and C. E. Forth, 'The Belly and Beyond: Body, Self and Culture in Ancient and Modern Times', in Carden-Coyne and Forth (eds), *Cultures of the Abdomen*, pp. 1–11.

— (eds), *Cultures of the Abdomen: Diet, Digestion and Fat in the Modern World* (New York: Palgrave Macmillan, 2005).

Cartwright, F. F., *The Development of Modern Surgery* (London: Arthur Baker Ltd, 1967).

Christie, D. A., and E. M. Tansey (eds), *Peptic Ulcer: Rise and Fall* (London: Wellcome Trust, 2002).

Clarke, E., and L. S. Jacyna, *Nineteenth-Century Origins of Neuroscientific Concepts* (Berkeley and Los Angeles, CA: University of California Press, 1987).

Clarkson, L. A., and E. M. Crawford, *Feast and Famine: A History of Food and Nutrition in Ireland 1500–1920* (Oxford: Oxford University Press, 2001).

Cleve, T. L., *The Saccharine Disease* (Bristol: John Wright, 1974).

Coker, W. J., B. M. Bhatt, N. J. Blatchley and J. T. Graham, 'The Clinical Findings of the first 1000 Gulf War Veterans in the Ministry of Defence's Medical Assessment Programme', *British Medical Journal*, 318 (1999), pp. 290–4.

Coleman, W., and F. L. Holmes (eds), *The Investigative Enterprise: Experimental Physiology in Nineteenth-Century Medicine* (Berkeley, CA: University of California Press, 1988).

Colley, L., *Britons: Forging the Nation 1707–1838* (New Haven, CT, and London: Yale University Press, 1992).

Collins English Dictionary, 6th edn (London: Harper Collins, 2003).

Condrau, F., and M. Worboys, 'Second Opinions: Epidemics and Infections in Nineteenth-Century Britain', *Social History of Medicine*, 20:1 (2007), pp. 147–58.

Cooper, C. L., *Stress: A Brief History* (Oxford: Blackwell, 2004).

Copp, J. T., and B. McAndrew, *Battle Exhaustion: Soldiers and Psychiatrists in the Canadian Army 1939–1945* (Montreal: McGill-Queen's University Press, 1990).

Crawford, E., *The Women's Suffrage Movement: A Reference Guide 1866–1928* (New York: Routledge, 2001).

Cunningham, P. A., *Reforming Women's Fashion 1850–1920* (Kent, OH: Kent State University Press, 2003).

Dally, A., *Women under the Knife: A History of Surgery* (London: Hutchison Radius, 1991).

—, *Fantasy Surgery 1880–1930 with Special Reference to Sir William Arbuthnot Lane* (Amsterdam: Rodopi, 1996).

Delaney, D. C., M. Qume et al., 'Helicobacter Pylori Test and Treat versus Proton Pump Inhibitor in Initial Management of Dyspepsia in Primary Care: Multicentre Randomised Controlled Trial', *British Medical Journal*, 336 (2008), pp. 651–4.

Desmond, A., *The Politics of Evolution: Morphology, Medicine and Reform in Radical London* (Chicago, IL, and London: Chicago University Press, 1989).

Digby, A., 'Women's Biological Straitjacket', in S. Mendus and J. Rendall (eds), *Sexuality and Subordination: Interdisciplinary Studies of Gender in the Nineteenth Century* (London and New York: Routledge, 1989), pp. 192–220.

Donaldson, F. L., *King George VI and Queen Elizabeth* (London: Weidenfield & Nicolson).

Dormandy, T., *The Fight Against Pain* (New Haven, CT: Yale University Press, 2006).

Drinka, G. F., *The Birth of Neurosis: Myth, Malady and the Victorians* (New York: Simon & Schuster, 1984).

Duchesneau, F., 'Vitalism in Late Eighteenth-Century Physiology: The Cases of Barthes, Blumenbach and John Hunter', in W. F. Bynum and R. Porter (eds), *William Hunter and the Eighteenth-Century Medical World* (Cambridge: Cambridge University Press, 1985), pp. 259–95.

Duffin, J., 'Laennec and Broussais: The "Sympathetic" Duel', in C. Hannaway and A. L. Berge (eds), *Constructing Paris Medicine* (Amsterdam: Rodopi, 1998), pp. 251–74.

Ellis, H., 'Some British Contributions to Abdominal Surgery 1870–1910', in Bynum (ed.), *Gastroenterology in Britain*, pp. 29–43.

Elston, M. A., 'Women and Anti-Vivisection in Victorian England, 1870–1900', in Rupke (ed.), *Vivisection in Historical Perspective*, pp. 259–94.

Engel, H., 'Billroth, Christian Albert Theodor', in C. C. Gillispie (ed.), *Dictionary of Scientific Biography 2* (New York: Charles Scribner's Sons, 1970), pp. 129–31.

Finlay, M. R., 'Quackery and Cookery: Justus von Liebig's Extract of Meat and the Theory of Nutrition in the Victorian Age', *Bulletin of the History of Medicine*, 66:3 (1992), pp. 404–18.

Fissell, M. E., 'The Disappearance of the Patient's Narrative and the Invention of Hospital Medicine', in French and Wear (eds), *British Medicine in an Age of Reform*, pp. 92–109.

Foucault, M., *The Birth of the Clinic* (Paris: Presses Universitaires de France, 1963).

Frank, R. G., *Harvey and the Oxford Physiologists: Scientific Ideas and Social Interaction* (Berkeley, CA, and London: University of California Press, 1980).

Frank, R. G. Jr, 'The Telltale Heart: Physiological Instruments, Graphic Methods and Clinical Hopes 1854–1914', in Coleman and Holmes (eds), *The Investigative Enterprise*, pp. 211–90.

French, D., *Raising Churchill's Army: The British Army and the War against Germany 1919–1945* (Oxford: Oxford University Press, 2000).

French, R. D., *Antivivisection and Medical Science in Victorian Society* (London and Princeton, NJ: Princeton University Press, 1976).

French, R. K., *Robert Whytt: The Soul and Medicine* (London: Wellcome Institute for the History of Medicine, 1969).

French, R. K., and A. Wear (eds), *British Medicine in an Age of Reform* (London and New York: Routledge, 1991).

Geddes, J. F., 'Culpable Complicity: The Medical Profession and the Forcible Feeding of Suffragettes, 1909–1914', *Women's History Review*, 17:1 (2008), pp. 79–94.

Geison, G. L., *Michael Foster and the Cambridge School of Physiology: The Scientific Enterprise in Late Victorian Society* (Princeton, NJ: Princeton University Press, 1978).

Gershon, M. D., *The Second Brain: A Groundbreaking New Understanding of Nervous Disorders of the Stomach and Intestine* (New York and London: Harper Collins, 1998).

Gibbs, D., 'The Demon of Dyspepsia: Some Nineteenth-Century Perceptions of Disordered Digestion', in Bynum (ed.), *Gastroenterology in Britain*, pp. 29–42.

Gilman, S. L., *Fat: A Cultural History of Obesity* (Cambridge: Polity Press, 2008).

Goodwin, C. S., 'Duodenal Ulcer, Campylobacter Pylori and the Leaking Roof Concept', *Lancet*, 332 (1988), pp. 467–9.

Grant, M., *Galen on Food and Diet* (London: Routledge, 2000).

Grieder, T., 'The German Drama in England, 1790–1800', *Restoration and Eighteenth-Century Theatre Research*, 32:2 (1964), pp. 39–50.

Grob, G. N., 'The Rise of Peptic Ulcer 1900–1950', *Perspectives in Biology and Medicine*, 46:4 (2003), pp. 550–66.

Gunn, S., *The Public Culture of the Victorian Middle Class* (Manchester and New York: Manchester University Press, 2000).

Haley, B., *The Healthy Body and Victorian Culture* (Cambridge, MA, and London: Harvard University Press, 1978).

Hall, P., *Cities in Civilisation: Culture, Innovation and Urban Order* (London: Weidenfield & Nicolson, 1998).

Hamilton, S. (ed.), *Animal Welfare and Anti-Vivisection 1870–1910: Nineteenth Century Women's Mission* (New York: Routledge, 2004).

Hamlin, C., *Public Health and Social Justice in the Age of Chadwick* (Cambridge: Cambridge University Press, 1998).

Hardy, A., 'The Medical Response to Epidemic Disease during the Long Eighteenth Century', in J. A. I. Champion (ed.), *Epidemic Disease in London*, Centre for Metropolitan History Working Papers Series, No. 1 (London: Centre for Metropolitan History, 1993), pp. 65–70.

—, 'Development of the Prison Medical Service 1774–1895', in R. Creese, W. F. Bynum and J. Bearn (eds), *The Health of Prisoners: Historical Essays* (Amsterdam: Rodopi, 1995), pp. 59–82.

Harrison, J. F. C., 'Early Victorian Radicals and the Medical Fringe', in W. F. Bynum and R. Porter (eds), *Medical Fringe and Medical Orthodoxy 1750–1850* (London: Croom Helm, 1987), pp. 198–215.

Harrison, M., *Medicine and Victory: British Military Medicine in the Second World War* (Oxford: Oxford University Press, 2004).

Heller, A., *A Theory of Modernity* (Malden, MA, and Oxford: Blackwell Publishers, 1999).

Henderson, J., *A Life of Ernest Starling* (New York and Oxford: Oxford University Press, 2005).

Hof, S., *Anorexia Nervosa: The Historical and Cultural Specificity: Fallacious Theories and Tenacious 'Facts'* (Lisse: Swets & Zeitlinger, 1994).

Holledge, J., *Innocent Flowers: Women in the Edwardian Theatre* (London: Virago Press, 1981).

Holmes, F. L., *Claude Bernard and Animal Chemistry: The Emergence of a Scientist* (Cambridge, MA: Harvard University Press, 1974).

—, 'The Physical Sciences in the Life Sciences', in M. J. Nye (ed.), *The Modern Physical and Mathematical Sciences* (Cambridge: Cambridge University Press, 2002), pp. 219–36.

Howarth, P., *George VI: A New Biography* (London: Hutchinson, 1987).

Howell, J. D., 'Soldier's Heart: The Redefinition of Heart Disease and Speciality Formation in Early Twentieth-Century Great Britain', *Medical History*, Supplement 5:1 (1985), pp. 1–33.

Huggett, J., *The Mirror of Health: Food, Diet and Medical Theory 1450–1660* (Bristol: Stuart Press, 1995).

Hughes, R. E., 'George Budd (1808–1882) and Nutritional Deficiency Diseases', *Medical History*, 17:2 (1973), pp. 127–35.

Jacyna, S., 'Images of John Hunter in the Nineteenth Century', *History of Science*, 21:1 (1983), pp. 85–108.

—, 'The Laboratory and the Clinic: The Impact of Pathology on Surgical Diagnosis in the Glasgow Western Infirmary 1875–1910', *Bulletin of the History of Medicine*, 62:3 (1988), pp. 384–406.

Jewson, N. D., 'The Disappearance of the Sick-Man from Medical Cosmology', *Sociology*, 10:2 (1976), pp. 225–44.

Jones, E., and S. Wessely, 'War Syndromes: The Impact of Culture on Medically Unexplained Symptoms', *Medical History*, 49:1 (2005), pp. 55–78.

Kinealy, C., *This Great Calamity: The Irish Famine 1845–1852* (Dublin: Gill & Macmillan, 1994).

King, L. S., *The Philosophy of Medicine: The Early Eighteenth Century* (London: Harvard University Press, 1978).

Korsmeyer, C., *Making Sense of Taste: Food and Philosophy* (Ithaca, NY: Cornell University Press, 1999).

Kurlansky, M., *Salt: A World History* (London: Vintage, 2003).

Lambert, W. R., *Drink and Sobriety in Victorian Wales 1820–1895* (Cardiff: University of Wales Press, 1983).

Lansbury, C., *The Old Brown Dog: Women, Workers and Vivisection in Edwardian England* (Madison, WI: University of Wisconsin Press, 1985).

Lau, W. Y., and C. K. Leow, 'History of the Perforated Duodenal and Gastric Ulcers', *World Journal of Surgery*, 21:8 (1997), pp. 890–6.

Lawrence, C., 'The Nervous System and Society in the Scottish Enlightenment', in B. Barnes and S. Shapin (eds), *Natural Order: Historical Studies of Scientific Culture* (London: Sage Publications, 1979), pp. 19–40.

—, 'Incommunicable Knowledge: Science, Technology and the Clinical Art in Britain', *Journal of Contemporary History*, 20:4 (1985), pp. 503–20.

—, 'Moderns and Ancients: The "New Cardiology" in Britain 1880–1930', *Medical History*, Supplement 5:1 (1985), pp. 1–33.

—, 'Democratic, Divine and Heroic: The History and Historiography of Surgery', in Lawrence (ed.), *Medical Theory, Surgical Practice*, pp. 1–47.

— (ed.), *Medical Theory, Surgical Practice: Studies in the History of Surgery* (London and New York: Routledge, 1992).

—, 'Medical Minds, Surgical Bodies: Corporeality and the Doctors', in C. Lawrence and S. Shapin (eds), *Science Incarnate: Historical Embodiments of Natural Knowledge* (Chicago, IL, and London: Chicago University Press, 1998), pp. 156–201.

Lawrence, C., and G. Weisz (eds), *Greater than the Parts: Holism in Biomedicine, 1920–1950* (Oxford: Oxford University Press, 1998).

Lederer, S. E., *Subjected to Science: Human Experimentation in America before the Second World War* (Baltimore, MD, and London: Johns Hopkins University Press, 1995).

Leese, P., *Shell Shock: Traumatic Neurosis and the British Soldiers of the First World War* (New York: Palgrave, 2002).

Lesch, J. E., *Science and Medicine in France: The Emergence of Experimental Physiology 1790–1855* (Baltimore, MD: Johns Hopkins University Press, 1967).

Levenstein, S., 'Stress and Peptic Ulcer: Life Beyond Helicobacter', *British Medical Journal*, 316 (1998), pp. 538–41.

—, '"The Very Model of a Modern Etiology": A Biopsychosocial View of Peptic Ulcer', *Psychosomatic Medicine*, 62:3 (2000), pp. 176–85.

Longmate, N., *The Waterdrinkers* (London: Hamish Hamilton, 1968).

Loudon, I., 'Medical Practitioners 1750–1850 and the Period of Medical Reform in Britain', in Wear (ed.) *Medicine in Society*, pp. 219–47.

Mackay, R., *Half the Battle: Civilian Morale in Britain during the Second World War* (Manchester: Manchester University Press, 2002).

Macrae, D. G., 'The Body and Social Metaphor', in J. Benthall and T. Polhemus (eds), *The Body as a Medium of Expression* (London: Allen Lane, 1975), pp. 59–73.

Maulitz, R., *Morbid Appearances: The Anatomy of Pathology in the Early Nineteenth Century* (Cambridge: Cambridge University Press, 1987).

May, M. T., *Galen on the Usefulness of the Parts of the Body* (Ithaca, NY: Cornell University Press, 1968).

Mazumdar, P., 'Anatomy, Physiology and Surgery: Physiology Teaching in Early Nineteenth-Century London', *Canadian Bulletin of Medical History*, 4:2 (1987), pp. 119–43.

Merriam-Webster's Medical Dictionary (Springfield, MA: Merriam-Webster Inc., 1995).

Morrell, J., *The Chemist Breeders: The Research Schools of Liebig and Thomas Thomson* (Cambridge: Heffer, 1972).

Morris, R. J., and R. Rodger, 'An Introduction to British Urban History 1820–1914', in R. J. Morris and R. Rodger (eds), *The Victorian City: A Reader in British Urban History 1820–1914* (London and New York: Longman, 1993), pp. 1–39.

Moyer, H. N., 'John Abercrombie: The First Neurologist', *Bulletin of the Society of Medical History of Chicago*, 2:3 (1920), pp. 137–45.

Mumford, L., *The City in History: Its Origins, its Transformations and its Prospects* (New York: Harcourt, Brace & World, 1961).

Normandin, S., 'Claude Bernard and an Introduction to the Study of Experimental Medicine: Physical Vitalism, Dialetic and Epistemology', *Journal of the History of Medicine and Allied Sciences*, 62:4 (2007), pp. 495–528.

Nunn, H. M., 'Home Bodies: Matters of Weight in Renaissance Women's Medical Manuals', in E. Klaver (ed.), *The Body in Medical Culture* (New York: Suny Press, 2009), pp. 15–36.

Ó Gráda, C., *Ireland before and after the Famine: Explorations in Economic History 1800–1925*, 2nd edn (Manchester and New York: Manchester University Press, 1993).

O'Neill, J., *Five Bodies: The Human Shape of Modern Society* (Ithaca, NY, and London: Cornell University Press, 1985).

Pagel, W., *Joan Baptista van Helmont: Reformer of Science and Medicine* (Cambridge: Cambridge University Press, 1982).

Pall, M., *Explaining 'Unexplained Illness'* (New York: Binghamton, 2007).

Pernick, M., *A Calculus of Suffering: Pain, Professionalism and Anaesthesia in Nineteenth-Century America* (New York: Columbia University Press, 1985).

Pitman, J., 'The John Abercrombie Collection', *Proceedings of the Royal College of Physicians of Edinburgh*, 21:3 (1991), pp. 349–54.

Poovey, M., *Making a Social Body: British Cultural Formation 1830–1864* (Chicago, IL, and London: Chicago University Press, 1995).

Porter, D. (ed.), *Social Medicine and Medical Sociology in the Twentieth Century* (Amsterdam: Rodopi, 1997).

Porter, R., 'Consumption: Disease of the Consumer Society?', in J. Brewer and R. Porter (eds), *Consumption and the World of Goods* (London and New York: Routledge, 1993), pp. 58–81.

—, 'Gout: Framing and Fantasizing Disease', *Bulletin of the History of Medicine*, 68:1 (1994), pp. 1–28.

—, 'Biliousness', in Bynum (ed.), *Gastroenterology in Britain*, pp. 7–28.

—, *Quacks: Fakers and Charlatans in English Medicine* (Stroud: Tempus, 2000).

—, *Bodies Politic: Disease, Death and Doctors in Britain 1650–1900* (London: Reaktion, 2003).

Porter, R., and G. S. Rousseau, *Gout: The Patrician's Malady* (New Haven, CT: Yale University Press, 1998).

Purvis, J., 'The Prison Experiences of the Suffragettes in Edwardian Britain', *Women's History Review*, 4:1 (1995), pp. 103–33.

Raban, J., *Soft City* (London: Fontana/Collins, 1974).

Richards, S., 'Drawing the Life-Blood of Physiology: Vivisection and the Physiologists' Dilemma, 1870–1900', *Annals of Science*, 43:11 (1986), pp. 27–56.

—, 'Vicarious Suffering, Necessary Pain: Physiological Method in Late Nineteenth-Century Britain', in Rupke (ed.), *Vivisection in Historical Perspective*, pp. 125–48.

Richardson, R., *Death, Dissection and the Destitute* (London: Routledge & Kegan Paul, 1987).

Rogers, A. I., and D. Hoel, 'Peptic Ulcer Disease: Retracing Science's Journey through the Gut', *Postgraduate Medicine*, 102:5 (1997), pp. 158–67.

Rosenberg, C. E., *Explaining Epidemics and other Studies in the History of Medicine* (Cambridge: Cambridge University Press, 1992).

Rosenberg, C. E., and J. Golden (eds), *Framing Disease: Studies in Cultural History* (New Brunswick, NJ: Rutgers University Press, 1992).

Rousseau, G. S., *Nervous Acts: Essays on Literature, Culture and Sensibility* (Basingstoke: Palgrave Macmillan, 2004).

Rupke, N. A. (ed.), *Vivisection in Historical Perspective* (London: Croom Helm, 1987).

Schoenfeldt, M., 'Fables of the Belly in Early Modern England', in D. Hillman and C. Mazzio (eds), *The Body in Parts: Fantasies of Corporeality in Early Modern Europe* (New York: Routledge, 1997), pp. 243–61.

Shapin, S., 'Trusting George Cheyne: Scientific Expertise, Common Sense, and Moral Authority in Early Eighteenth-Century Dietetic Medicine', *Bulletin of the History of Medicine*, 77:2 (2003), pp. 263–97.

Shepherd, B., *A War of Nerves: Soldiers and Psychiatrists 1914–1994* (Cambridge, MA: Harvard University Press, 2001).

Shorter, E., 'The First Great Increase in Anorexia Nervosa', *Journal of Social History*, 21:1 (1987), pp. 69–96.

—, *Women's Bodies: A Social History of Women's Encounters with Health, Ill-Health and Medicine* (New Brunswick, NJ: Transaction Publishers, 1991).

—, *From Paralysis to Fatigue: A History of Modern Psychosomatic Medicine* (New York: Free Press, 1992).

—, *From the Mind into the Body: The Cultural Origins of Psychosomatic Symptoms* (New York: Free Press, 1994).

Siegel, R. E., *Galen's System of Physiology and Medicine* (Basel: Karger, 1968).

Sim, J., *Medical Power in Prisons: Prison Medical Service in England 1774–1988* (Buckingham: Open University Press, 1990).

Smith, D. C., 'Appendicitis, Appendectomy and the Surgeon', *Bulletin of the History of Medicine*, 70:3 (1996), pp. 414–41.

Smith, G. D., 'The Biopsychosocial Approach: A Note of Caution', in P. White (ed.), *Biopsychosocial Medicine: An Integrated Approach to Understanding Illness* (Oxford: Oxford University Press, 2005), pp. 77–103.

Smith, R., *Trial by Medicine: Insanity and Responsibility in Victorian Trials* (Edinburgh: Edinburgh University Press, 1981).

Smith Kline and French International Co., *The H2 Receptor Anthology: The Discovery of Histamine H2-Receptors and their Antagonists* (Welwyn Garden City: Smith Kline & French, 1982).

Snow, S. J., *Operations without Pain: The Practice and Science of Anaesthesia in Victorian Britain* (Basingstoke: Palgrave Macmillan, 2006).

Sonnenberg, A., 'Causes Underlying the Birth-Cohort Phenomenon of Peptic Ulcer: Analysis of Mortality Data 1911–2000, England and Wales', *International Journal of Epidemiology*, 35:4 (2006), pp. 1090–7.

Spadafora, D., *The Idea of Progress in Eighteenth-Century Britain* (New Haven, CT, and London: Yale University Press, 1990).

Spiro, H., 'Peptic Ulcer is not a Disease, Only a Sign! Stress is a Factor in More than a Few Dyspeptics', *Psychosomatic Medicine*, 62:2 (2000), pp. 186–7.

Stanley, P., *For Fear of Pain: British Surgery 1790–1850* (Amsterdam and New York: Rodopi, 2003).

Stevenson, R. S., *Famous Illnesses in History* (London: Eyre & Spottiswoode, 1962).

Susser, M., 'Period Effects, Generation Effects and Age Effects in Peptic Ulcer Mortality', *Journal of Chronic Disease*, 35:1 (1982), pp. 29–40.

Temkin, O., 'The Scientific Approach to Disease: Specific Entity and Individual Sickness', in A. C. Crombie (ed.), *Scientific Change: Historical Studies in the Intellectual, Social and Technical Conditions for Scientific Discovery and Technical Innovation from Antiquity to the Present* (New York: Basic Books, 1963), pp. 629–47.

Thagard, P., *How Scientists Explain Disease* (Princeton, NJ: Princeton University Press, 1999).

Thalassis, N., 'The Use of Intelligence Testing in the Recruitment of "Other Ranks" in the Armed Forces During the Second World War', *History of Philosophy and Psychology*, 5 (2003), pp. 17–29.

Thornton, J. L., *John Abernethy: A Biography* (London: Simpkin Marshall, 1953).

Timmermann, C., 'Constitutional Medicine, Neoromanticism, and the Politics of Antimechanism in Interwar Germany', *Bulletin of the History of Medicine*, 75:1 (2001), pp. 27–39.

Todes, D. P., *Pavlov's Physiology Factory: Experiment, Interpretation, Laboratory Enterprise* (Baltimore, MD: Johns Hopkins University Press, 2002).

Townend, P., *Father Mathew, Temperance and Irish Identity* (Dublin: Irish Academic Press, 1992).

Tracy, S. W., 'George Draper and American Constitutional Medicine 1916–1946: Re-Inventing the Sick Man', *Bulletin of the History of Medicine*, 66:1 (1992), pp. 53–89.

Tröhler, U., 'To Operate or Not to Operate: Scientific and Extraneous Factors in Therapeutical Controversies within the Swiss Society of Surgery 1913–1988', in W. F. Bynum and V. Nutton (eds), *Essays in the History of Therapeutics* (Amsterdam: Rodopi, 1991), pp. 89–114.

Trombley, S., *Sir Frederick Treves: The Extra-ordinary Edwardian* (Routledge: London, 1989).

Turner, B. S., 'The Government of the Body, Medical Regimen and the Rationalisation of Diet', *British Journal of Sociology*, 33:2 (1982), pp. 254–69.

Vila, A. C., 'The Philosophe's Stomach: Hedonism, Hypochondria and the Intellectual in Enlightenment France', in Carden-Coyne and Forth (eds), *Cultures of the Abdomen*, pp. 89–104.

Waddington, I., *The Medical Profession in the Industrial Revolution* (Dublin: Gill & Macmillan, 1984).

Walkowitz, J. R., *Prostitution and Victorian Society: Women, Class and the State* (New York and Cambridge: Cambridge University Press, 1982).

Wangensteen, O. H., and S. D. Wangensteen, *The Rise of Surgery from Empiric Craft to Scientific Discipline* (Folkestone: Dawson, 1978).

Warner, J. H., 'Ideals of Science and their Discontents in Late Nineteenth-Century American Medicine', *Isis*, 82:3 (1991), pp. 454–78.

—, 'The Fall and Rise of Professionalism: Epistemology, Authority and the Emergence of Laboratory Medicine in Nineteenth-Century America', in A. Cunningham and P. Williams (eds), *The Laboratory Revolution in Medicine* (Cambridge: Cambridge University Press, 1992), pp. 110–41.

Warren, J. R., 'Unidentified Curved Bacilli on Gastric Epithelium in Active Chronic Gastritis', *Lancet*, 321 (1983), pp. 1273–5.

Watkins, E. S., 'Medicine, Masculinity and the Disappearance of Male Menopause in the 1950s', *Social History of Medicine*, 21:2 (2008), pp. 329–44.

Watson, J. R., 'John Abercrombie and the Diagnosis of Duodenal Ulcer', *Annals of Medical History*, 3:4 (1942), pp. 468–72.

Wear, A., 'Making Sense of Health and the Environment in Early Modern England', in Wear (ed.) *Medicine in Society*, pp. 119–47.

— (ed.), *Medicine in Society* (Cambridge, Cambridge University Press, 1992).

—, *Knowledge and Practice in English Medicine: 1550–1680* (Cambridge: Cambridge University Press, 2000).

Weineck, S. M., 'Digesting the Nineteenth Century: Nietzsche and the Stomach of Modernity', *Romanticism*, 12:1 (2006), pp. 35–43.

Wheelwright, J., *The Irritable Heart: The Medical Mystery of the Gulf War* (New York and London: Norton, 2001).

White, P., 'Sympathy under the Knife: Experimentation and Emotion in Late Victorian Medicine', in F. B. Alberti (ed.), *Medicine, Emotion and Disease 1700–1950* (Basingstoke: Palgrave Macmillan, 2006), pp. 100–24.

Whorton, J. C., *Inner Hygiene: Constipation and the Pursuit of Health in Modern Society* (Oxford: Oxford University Press, 2000).

Wilde, S., and G. Hurst, 'Learning from Mistakes: Early Twentieth-Century Surgical Practice', *Journal of the History of Medicine and Allied Sciences*, 64:1 (2009), pp. 38–77.

Williams, E. A., 'Neuroses of the Stomach: Eating, Gender, and Psychopathology in French Medicine 1800–1870', *Isis*, 98:11 (2007), pp. 54–97.

—, 'Gags, Funnels and Tubes: Forced Feeding of the Insane and of Suffragettes', *Endeavour*, 32:4 (2008), pp. 134–40.

Wilson, A., 'On the History of Disease-Concepts: The Case of Pleurisy', *History of Science*, 38:3 (2000), pp. 271–319.

Wolfe, R. J., *Tarnished Idol: William Thomas Green Morton and the Introduction of Surgical Anaesthesia: A Chronicle of the Ether Controversy* (San Anselmo, CA: Norman Publishing, 2001).

Wooley, C. F., *The Irritable Heart of Soldiers and the Origins of Anglo-American Cardiology* (Aldershot: Ashgate, 2002).

Worboys, M., *Spreading Germs: Diseases, Theories and Medical Practice in Britain 1865–1900* (Cambridge and New York: Cambridge University Press, 2000).

Wrigley, E. A., *Continuity and Change: The Character of the Industrial Revolution in England* (Cambridge: Cambridge University Press, 1988).

Young, A., *The Harmony of Illusions: Inventing Post-Traumatic Stress Disorder* (Princeton, NJ: Princeton University Press, 1995).

Zanten, L., and V. Zanten, 'The Aging Stomach or the Stomachs of the Ages', *Gut*, 41:4 (1997), pp. 575–6.

Zimmerman, L. M., and I. Veith, *Great Ideas in the History of Surgery* (1961; Novato, CA: Norman Publishing, 1993).

INDEX

www.ingramcontent.com/pod-product-compliance
Ingram Content Group UK Ltd.
Pitfield, Milton Keynes, MK11 3LW, UK
UKHW011012180926
475701UK00004B/694
* 9 7 8 1 1 3 8 6 6 4 5 0 0 *